AF805348

Emerging Trends in Mechatronics

Series Editor

Aydin Azizi, Oxford, UK

This book series helps understand how Mechatronics will impact the practice and research of developing advanced techniques to model, control and optimize complex systems. It presents recent state of advances in mechatronics and related technologies including: Automatic Control, Robotics, Agent Based systems, Smart Manufacturing and Industry 4.0. The selected topics give an overview of the state-of-the-art and present new research results and prospects of the future development in the interdisciplinary field of mechatronic systems. The book series will provide up-to-date and useful knowledge for researchers and engineers involved in mechatronics and related fields.

Within the above dimensions, the scope of the book series welcomes high quality original research and review articles that cover a broad range of topics related to theoretical and applied aspects of modern mechatronics. Potential topics include, but are not limited to the following:

Controller design

Many mechatronic systems have challenging control difficulties due to the high nonlinear structures and time-varying dynamic behaviors. In addition to these, there are external disturbances, which cannot be predicted and change according to uncontrolled working environments. Conventional controllers are insufficient to solve the problems and to compensate the environmental disturbances, therefore designing efficient controllers for complex systems has a great importance in the current book series.

Sensor Design & Data Collection Approaches

Essential part of a mechatronics system is the measurement system that senses the variations in the physical parameters, such as temperature, pressure, displacement, and so on, and converts it to voltage or current. In mechatronic systems, accurate measurement of the dynamic variables plays a vital role for the actuators to function properly, therefore on of the aims of this book series, to focus on understanding what should be done to design accurate sensors and efficient data collection methods to improve the performance of mechatronic systems.

Optimization Techniques

Designing a complex mechatronic system involves multiple variables, objectives, constraints, and evaluation criteria as well as their nonlinearly coupled relationships. One of the challenging research areas of this book series is utilizing optimization techniques to design an optimal mechatronic system.

Artificial Intelligence

From the point of view of dynamics, mechatronic systems can be characterized by model uncertainties, high nonlinearities, complicated coupling, and stringent performance requirements, among others. To deal with these facts, this book series would like to focus on investigating and utilizing Artificial Intelligence techniques such as Neural Networks, Fuzzy logic, Neuro-fuzzy, evolutionary techniques etc. to model, design and optimize complex mechatronic systems.

Robotics

In the field of Robotics, there is a need of focusing on kinematic, dynamic, control and optimization of innovative robotic systems. This book series can introduce new and novel robot designs such as hybrid (serial-parallel) redundant manipulators that have a high potential of application in special fields like space robots. State of the art researches on selection of optimum kinematic solutions and path planning based on such an optimization, dynamic analysis and new control methods developments and application of artificial intelligence in this field can be published in this book series.

Smart manufacturing

Application of networked sensors and machinery equipment using internet and intranet in order to monitor and automation of the manufacturing process and improve the performance efficiency and is a point of interest. Advanced robotics, multi input-multi output control approaches combined with the artificial intelligence and optimization techniques, big data processing approaches, multi-scale modelling and simulation of dynamic sub-systems etc. are the hot topics that will be focused in this book series.

AGVs

Automated Guided Vehicle (AGVs) are going to be very popular is industry logistics, surveillance, defense and social applications in order to make processes safer and more cost efficient. Innovative AGVs designs for various missions, Robot Operating System usage on AGV control, Indoor and outdoor navigation and control algorithms, intelligent path planning techniques, sensor fusion, simultaneous application of GPS, INS and offline maps are the topics that are planned to be published in this book series.

Aydin Azizi
Editor

Hybrid Energy Systems and AI Integration

Springer

Editor
Aydin Azizi
School of Engineering, Computing and Mathematics
Oxford Brookes University
Oxford, UK

ISSN 2731-4855 ISSN 2731-4863 (electronic)
Emerging Trends in Mechatronics
ISBN 978-981-95-5858-2 ISBN 978-981-95-5859-9 (eBook)
https://doi.org/10.1007/978-981-95-5859-9

© The Editor(s) (if applicable) and The Author(s), under exclusive license to Springer Nature Singapore Pte Ltd. 2026

This work is subject to copyright. All rights are solely and exclusively licensed by the Publisher, whether the whole or part of the material is concerned, specifically the rights of translation, reprinting, reuse of illustrations, recitation, broadcasting, reproduction on microfilms or in any other physical way, and transmission or information storage and retrieval, electronic adaptation, computer software, or by similar or dissimilar methodology now known or hereafter developed.
The use of general descriptive names, registered names, trademarks, service marks, etc. in this publication does not imply, even in the absence of a specific statement, that such names are exempt from the relevant protective laws and regulations and therefore free for general use.
The publisher, the authors and the editors are safe to assume that the advice and information in this book are believed to be true and accurate at the date of publication. Neither the publisher nor the authors or the editors give a warranty, expressed or implied, with respect to the material contained herein or for any errors or omissions that may have been made. The publisher remains neutral with regard to jurisdictional claims in published maps and institutional affiliations.

This Springer imprint is published by the registered company Springer Nature Singapore Pte Ltd.
The registered company address is: 152 Beach Road, #21-01/04 Gateway East, Singapore 189721, Singapore

If disposing of this product, please recycle the paper.

Preface

The rapid evolution of energy systems, coupled with the increasing complexity of industrial processes, necessitates the development of sophisticated diagnostic and monitoring tools. This book addresses this need by bringing together pioneering work from diverse but interrelated fields.

From the optimization of hybrid photovoltaic systems to the application of artificial intelligence in vibration-based fault detection, the chapters in this volume offer cutting-edge solutions and methodologies. Each contribution represents a significant advancement in its field, supported by rigorous research and validated by experimental or simulation results.

It is our hope that this book will serve as both a reference and inspiration for future research, bridging the gap between energy technology and diagnostics science.

Oxford, UK　　　　Aydin Azizi

About This Book

This book focuses on the convergence of hybrid energy systems and advanced diagnostics techniques, with a special emphasis on renewable energy integration and intelligent fault detection in complex engineering systems. It brings together innovative research contributions from leading institutions, offering readers comprehensive insights into both the theoretical foundations and practical applications of advanced diagnostics in energy and mechanical systems.

The topics covered range from photovoltaic system diagnostics to intelligent monitoring of industrial pumps and pipelines, providing a multi-disciplinary perspective that will be invaluable for engineers, researchers, and industry practitioners.

Contents

Advanced Diagnostics Methods for Hybrid Photovoltaic Systems

Ali Isam Mahmood Hajer and Mohammad Karimzadeh Kolamroudi

Abstract Stability of the power converters is therefore crucial in the effective performance of stand-alone hybrid PV systems. This research provides a diagnostic framework that is specifically designed for power converters in such systems. The new approach is based on the implementation of enhanced monitoring methods, fault diagnostics, and prognosis of maintenance actions to improve the system dependability and performance. The offered diagnostic framework is tested under different operating conditions and fault cases based on MATLAB for the investigation of diagnostic framework performance. The results obtained by the fault detection algorithms are quite noticeable. The statistical deviation method yielded an FDR of 92%, FAR of 5%, MTTD of 17 min, and a mean time to repair of 22 min. The performance of model-based diagnosis was evaluated in terms of FDR, FAR, MTTD, and MTTR, and it was recorded as 88%, 7%, 21 min, and 25 min, respectively. The outcomes emphasize the significance of high efficiency in the identification of the fault in enhancing the performance and dependability of standalone hybrid PV systems. The finally designed fault detection algorithms are characterized by high FDR, which, in turn, would ensure the correct identification of the faults and the reliability of the system, while low FAR values would point out the effectiveness of these algorithms in detecting the real anomalies.

Keywords Advanced diagnostics methods · Power converters · Standalone hybrid photovoltaic systems · MATLAB

A. I. M. Hajer
Department of Electrical and Electronics, Near East University, Nicosia, Türkiye

M. Karimzadeh Kolamroudi (✉)
Department of Mechanical Engineering, Near East University, Nicosia, Türkiye

Research Center for Science, Technology and Engineering (BILTEM), Near East University, Nicosia, Türkiye
e-mail: Mohammad.karimzadeh@neu.edu.tr

© The Author(s), under exclusive license to Springer Nature Singapore Pte Ltd. 2026
A. Azizi (ed.), *Hybrid Energy Systems and AI Integration*, Emerging Trends in Mechatronics, https://doi.org/10.1007/978-981-95-5859-9_1

1 Section I

1.1 Introduction

1.1.1 Importance of Energy

Clean and reliable energy sources are crucial for environmental and economic development, and standalone hybrid photovoltaic systems play a significant role in this. Industries, residences, and essential services rely on energy, and its sustainable generation is vital to the global fight against climate change. Stand-alone hybrid photovoltaic systems, which are clean, renewable, and often cheaper than fossil fuels, have advanced this effort. These systems reduce carbon emissions and supply electricity in rural or off-grid regions, thereby improving quality of life and promoting sustainable development [1, 2].

In such systems, improved power converter diagnostics are essential to maximize energy output and system longevity, enabling efficient solar energy harvesting. This combination of sustainable energy solutions and modern technology demonstrates a commitment to meeting energy needs in an environmentally, fiscally, and socially responsible way [3–5].

1.1.2 Motivation of the Study

The critical factors and criteria of sustainable energy systems inspired the paper "Advanced Diagnostics Methods for Power Converters in Stand-alone Hybrid Photovoltaic Systems." Summarizing the motivators:

Due to global efforts to reduce carbon emissions and mitigate climate change, renewable energy sources are becoming more critical. Solar electricity and other renewables are widespread. Standalone hybrid photovoltaic systems can generate energy in rural or off-grid areas. The study aims to help achieve a sustainable transition to renewable energy sources.

Hybrid photovoltaic systems need power converters to convert and regulate solar energy efficiently [6]. They are vital to system stability and reliability. However, converter failures and inefficiencies can reduce the photovoltaic system's efficiency.

Power converter diagnostics primarily rely on traditional procedures; hence, advanced methods are needed. Modern stand-alone hybrid photovoltaic systems have complex dynamics that traditional approaches may not handle. Diagnostic, efficient, precise methods that can predict defects are in demand. This work develops or improves diagnostic procedures for these systems to meet this gap.

The project emphasizes improved diagnostics to increase photovoltaic power converter reliability and efficacy. Constant power supply, fewer maintenance costs, and longer system component lifespan are crucial.

Improving power converter diagnostics and efficiency can decrease electricity generation and maintenance expenses. Optimizing renewable energy systems maximizes clean energy use and reduces fossil fuel use, promoting environmental sustainability.

The study aims to enhance renewable energy technology through innovation and technology. Diagnostic advances can spur technological discoveries and set a bar for future study.

This study was inspired by the goals of global energy security and accessibility. It improves the reliability and efficiency of independent hybrid photovoltaic systems, making renewable energy more accessible and feasible, especially in remote or underdeveloped locations, and promoting energy equity.

This project aims to improve stand-alone hybrid PV system performance and dependability to progress renewable energy. This will be done with advanced power converter diagnostics. The research affects the environment, economics, technology, and society [6].

1.1.3 Standalone Hybrid Photovoltaic (PV) Systems

The standalone hybrid photovoltaic (PV) systems are innovative solar energy technologies and solve modern energy production problems. These systems synergistically optimize energy availability and reliability by combining solar panels with wind turbines or diesel engines. This hybridization is an advantage over typical solar systems, which rely solely on the sun and are limited by weather and daily sessions [6]. Stand-alone hybrid PV systems compensate solar energy's intermittent nature with different energy sources to provide continuous power. Continuous power supply guarantees reliable power output, making these systems efficient [6].

Rural and off-grid locations benefit from standalone hybrid PV systems because they address a global energy distribution issue: unreliable and unreasonable electricity [6]. In many locations, especially in the remote areas, the concept of the electricity grid is completely unknown, and, even if was ever present, it has been discontinued rather often. These scenarios are ideal for isolated, operating on solar power only, hybrid PV systems that represent sustainability and self-sufficiency. They are very real and sustainable solutions to the grid-based electric supply by offering a renewable sourcing solution to the regions without direct access to the grid based energy sourcing. Generation of power from sources through a common grid enhances the energy autonomy and the protection of people against electrical blackouts and other structural problems.

The basic requirements of hybrid PV systems can be flexible according to the variety of environmental factors as well as the energy requirements apart from the operational utility. Focusing on the local climate, the availability of renewable resources, and the directions of the energy request, these regulations can be dedicated to operate the local energy needs. Stand-alone hybrid systems are applied due to their versatility in rustic homes and also the client's isolated industrial buildings. They also help in saving on the utilization of fossil fuels and hence they are environmentally friendly having embraced different sorts of renewable energy. Finally, the specifically installed hybrid PV systems prove the technological improvement of renewable energy and contribute to environmental sustainability and reduction of carbon marks. As this study shows standalone hybrid PV systems have the flexibility to meet sustainability and are beneficial to developing a sustainably robust global energy sector [6].

1.1.4 The Critical Role of Power Converters

As for the standalone hybrid PV systems, the power converters are basic and indispensable parts of the complex system because of their contribution to the energy conversion. These converters transform the direct current (DC) production into a alternating current (AC), which is required for domestic and industrial purposes. This is not a direct conversion; however, it includes a complex control and regulation procedure to guarantee the generated output to be in tune with the correct energy demand and grid condition. This paper established that the type of power converters, their efficiency and the type of resolution all influence the PV system efficiency [6]. They are involved on the complex operation of these systems to integrate solar energy into useful and functional systems. Power converters are also overseeing conversion from AC to DC which is important for storing energy in batteries. This goes a long way in showing their versatility when it comes to administering the energy of PV system.

Thus, power converters have crucial positions in converting the energy and upgrading the performance and efficiency of the PV systems. High efficiency power converters harvest maximum possible energy from solar panels and ensure that, solar energy is as easily harvested as possible for use by mankind. It also optimizes efficiency of converting input into output in the case of the energy controller and calls for even greater control regarding the energy flow in the system. Power converters are useful in sustaining the energy balance in hybrid systems since they interface with multiple power sources [7]. This is especially so where applications are off-grid and depend solely on such means for maximum reliability. furthermore, advancements in power converter technology, including the smart controls and diagnostics fascinatingly are able to increase the efficiency of the PV systems to a much higher level than before. This can lead to finer energy control, the appropriateness of which to expect maintenance requirements, and enhanced system longevity; this demonstrates that converters play a central role in advancing solar energy systems [6].

1.1.5 Need for Advanced Diagnostic Methods

For standalone hybrid photovoltaic systems, the power converters convert the energy, being a significant component of the system framework. Solar panels generate varying direct current (DC) and these converters help in converting the DC into correct AC for domestic and commercial use. There is a need to control and modulate to achieve the output conforming to the energy need and even the grid standards. Therefore, precision of PV systems efficiency depends on power converter efficiency. These systems depend on them to connect the use of solar energy into practical as well as functional solutions. The power converters change the AC for DC to store in batteries. This accounts for the effectiveness of the STS in managing the PV energy, regardless of the specific area of operation [8].

Power converters are necessary for energy conversion and PV system efficiency. Efficient power converters maximize solar energy utilized. This optimization improves conversion efficiency and requires discreet energy flow control [8]. Off-grid scenarios entail reliability; so, this is crucial. Improving the power converter technology and also can boost PV system efficiency, including intelligent controls and diagnostics. Power converters are vital to solar energy system upgrade due they can increase energy control, maintenance prediction, and system durability [8].

1.1.6 Long-term Sustainability

The headway in diagnostic techniques for power converters goes behind a simple technical improvement; it posed a significant step toward ensuring the long-term viability of hybrid photovoltaic (PV) systems. The consequences of incorporating enhanced diagnostics into standalone hybrid PV systems are important and extensive. trusted renewable energy systems play a crucial role in the regard of global environmental safeguard and the pursued of sustainable development.

Enhanced diagnosis ensures that such systems operate at the highest optimization level and hence maximize the utilization of renewable material and minimize the amount of waste that is produced [9]. The effectiveness of this operation is important in reducing the carbon activity with regard to energy production since it has a natural allegiance in similar efforts being carried out globally to conserve the environment and reduce greenhouse emissions. Thus, the application of complex diagnostic methods in power converters contributes to the creation of more efficient renewable energy systems.

1.1.7 Problem Statement

Thus, the conversion of a standalone hybrid photovoltaic (PV) system is adequately effective provided that the efficiency of the power converters is paramount. But these converters are not devoid of some faults and inefficiencies, which are always disadvantageous as far as the entire system is concerned. There is no much development of sophisticated diagnostic method aimed at driving converters in standalone hybrid PV systems which held back early identification and relief of problems [8].

The system lacks the diagnostic algorithms capable of detecting and analyzing defects in power converters, which reduce overall system reliability and performance. It is important to fill this gap in order to control and maintain standalone hybrid PV systems efficacies and durability in the long-run. In order to achieve these objectives, the use of elaborate diagnostic strategies that include equations for converter efficiency, power quality index and total harmonic distortion in the drafting and application procedures seeks to enhance power converters' reliability and thus, the performance of standalone hybrid PV systems [10].

1.1.8 Rational of the Study

Solving a Crucial Requirement in Renewable Energy Systems

The stand-alone hybrid PV systems are very viable and are also reliable sources of energy for the long term since the world is shifting to the use of solar electricity. Each of these systems is highly reliant on the efficiency, and sometimes even the reliability, of the performance of power converters. Power converters use solar energy as usable power, yet, component failures, efficiency losses, and foreign factors may be realigning it. These barriers probably affect the efficiency and the longevity of photovoltaic (PV) systems. Therefore, early diagnostic and preventive mechanisms for these problems are required to maintain these systems' smooth and efficient functioning.

1.1.9 Addressing the Discrepancy in Existing Studies

There is literature review on the evaluation of power converters in photovoltaic (PV) systems but scanty literature on improved diagnostic techniques. According to conventional diagnostic procedures, most of the time deal with a definite fault after it incapacitated the system. This work examines preemptive diagnostic tools like machine learning techniques and big data analysis to plug this hole and enhance problems detection and prevent. The case brings fresh concepts and approaches to an exhausted renewable energy technology object.

1.1.10 Conforming to International Sustainability Objectives

This research is justified on the grounds of being in line with the planned sustainable development goals on the international level. The global move to cut on the use of fossil fuel and greenhouse gases call for efficiency in generation of renewable power. Advanced and accurate diagnosis of power converters can extend the life cycle and energy efficiency of photovoltaic systems which consequently enhances the energy sector's sustainability and ecological advantage. This problem fits perfectly into the technological category, as well as into the category of ecological necessity.

1.1.11 Promoting Technological Advancement

Recently, the study offered advances in sustainable energy technology. Renewable energy and data science meet in power converter diagnostics with both machine learning and predictive analytics. This integration can set a new standard for renewable energy system performance and maintenance, enabling greater intelligent, efficient, and robust energy systems.

1.1.12 Research Question and Hypothesis

The Following Questions and Hypotheses Guide of this Research

Research Q 1: What are the typical issues in power converters using in standalone hybrid photovoltaic systems, and how can they be fidelity detected using advanced diagnostic techniques?

Hypothesis. The primary concerns indicated in converters for standalone hybrid photovoltaic systems concern thermal effect, electrical stressing, and degradation of components. Thermal imaging and vibration analysis are some of the most effective ways of identifying these problems due their effectiveness in diagnosis. With regard to Big Data, it can accurately detect all the tomfoolery and oddities of operating data.

Research Q 2: Can we monitor the power converters in real-time using advanced diagnostic methods result in observable enhancements in the operational efficiency of standalone hybrid photovoltaic systems?

Hypothesis. Implementing advanced diagnostic methods allows real-time monitoring of power converters in standalone hybrid photovoltaic systems, substantially improving operational efficiency. These benefits result from identifying issues early on, streamlining power conversion processes, and minimizing upkeep outage.

Research Q 3: How do environmental factors affect the power converters' performance and failure rates in standalone hybrid PV systems? Additionally, how could diagnostic methods be customized to consider these factors?

Hypothesis. Various parameters such as temperature, humidity, and dust affect the efficiency as well as the reliability of power converters in standalone hybrid PV system. Thus, by working in combinative with adjustment algorithms and environmental sensors, diagnostic methods can be marked as taking into consideration the environmental factors. It can greatly improve the accuracy and speed of flaw identification and instrumental system repair and maintenance.

1.1.13 Novelty of the Research

This revolutionary methodology presents a distinctive new step that comprises new state of the art techniques for diagnosing power converters in standalone hybrid PV systems. Factors that shareholding significantly to originality include:

1.1.14 Integration of Multiple Renewable Sources

The combination of PV and wind power is an efficient system of creating energy in tandem with the use of wind turbines. This strategy helps to obtain more reliable and effective power from the multiple resources.

1.1.15 Advanced Fault Detection Algorithms

The main key methods used in this work are fault detection, diagnosis and improvement including machine learning, statistical deviation analysis and kalman filtering. These techniques are expected to enhance the system's ability to detect, predict and handle faults in real time.

1.1.16 Comprehensive Performance Evaluation Metrics

Between the performances evaluation criteria established in the study are the following:

The key performance indicators of a fault management program include Fault Detection Rate (FDR), False Alarm Rate (FAR), Mean Time to Detect (MTTD), and Mean Time to Repair (MTTR). Necessary and sufficient conditions are provided by this wide set of standards, and the authenticity and conduct of the system are proving.

1.1.17 Significance of the Study

It can therefore be described as a potentially groundbreaking study because if the identified problems prove to be major, their solutions can benefit other hybrids as well. Consequently, this research is valuable for the development of renewable energy technology, thereby touching on general environmental concern, economic efficiency, and energy security.

1.1.18 Progress in Renewable Energy Technology

Whether it be through direct investment or installing project facilities, this work's connection with renewable energy technology is important. Regarding the objectives of the study, power converter diagnostics is a key area of interest since the ability to accurately diagnose the power converter is a major challenge faced when it comes to expanding the efficiency and reliability of standalone hybrid PV systems. These are useful systems for the shift to renewable energy sources, and efficiency improvements immediately reflect in the solar power sector. The research focuses on the approaches like predictive analytics and machine learning that might be useful in problem finding as well as system maintenance. Thanks to this, the power and durability of photovoltaic systems will be brought to the highest level, thus expanding the applicability and popularity of the solar energy project.

1.1.19 Ecological Consequences and Long-Term Viability

In the study, these preventive measures were presented under the headings of predictive analytics and machine learning as approaches to troubleshooting and maintenance. This discovery will lead to a longer lifespan as well as improved efficiency levels for photovoltaic systems and, therefore, make solar energy a more suitable and desirable option for a broader number of uses.

1.1.20 Contribution to the Advancement of Research and Formulation of Policies

This work matters to academics and politicians. It provides vital knowledge to renewable energy technology, which may inform future study. The findings can inform energy policy, particularly in promoting sustainable sources and infrastructure. This research can help policymakers develop strategies and frameworks to encourage renewable energy systems to use advanced diagnostic tools, furthering sustainable energy regulations.

1.1.21 Structure of the Research

This research organizational framework comprises numerous parts; the second is an exhaustive literature review that establishes the theoretical groundwork for the investigation. In subsequent chapters, the technique is broken down in further detail, the findings are presented, and the consequences of the findings are discussed. Conclusions and suggestions for further research in this area are included in the final chapter of the study.

2 Section II

2.1 Literature Review

2.1.1 Introduction

This research demonstrates that the integration of advanced diagnostic techniques for power converters is crucial to improving the reliability, operational continuity, and durability of standalone hybrid PV systems. The literature search regarding diagnostic procedures that can subsume power converters in hybrid PV systems starts with this chapter. Reliability and efficacy of power converters are major determining factors when it comes to optimizing stand-alone hybrid systems which are essential with growing demand of renewable and self-sufficient energies.

Among the various potential sources of energy other than the conventional sources of power, one of the interactive options that are proving popular is the stand-alone hybrid PV system. Combining it with other components that either store or create a greater amount of energy, the solutions based on photovoltaic technology allow for providing a durable, and long-term supply of electric power. However, for these systems to function properly, power converters have to be healthy and performing to the utmost, and therefore, advanced diagnosis methods are required [6].

2.1.2 Related work

Optimized Diagnostic Techniques for Power Converters in Standalone Hybrid Wind Energy Systems

Independent Hybrid wind energy systems are one of the modern developments in the renewable energy field for off-grid and remote areas. It emphasizes the power converter as a very important element that helps to design one of the most efficient and, at the same time, reliable power supply systems. It is these converters' role to stabilize the flow of wind energy since it cannot be constant like the energy produced by a wind turbine [11]. These converters should be capable of responding to dynamics of wind speed and loads, for which optimization algorithm used in the design stage. This makes it possible to minimize the amount energy that is wasted and also increase efficiency of the system. The application of such efficient power converters additionally boosts the significance of such converters when integrated into hybrid systems that combine wind, where the latter often combines with other renewable sources such as solar energy [12]. Not only they make it easier to incorporate the various sources of power supply into the complete power supply system without much of issue but they also ensure an improved and sustainable form of power supply especially for those regions which are not privileged to have the power grid. Therefore, the synthesis and implementation of these optimized power converters are critical to the FHEHSs' popularity and effectiveness in standalone hybrid wind power systems [13]. This is a big leap in the quest for renewable energy sources as illustrated in Fig. 1 where stand-alone hybrid wind and solar energy arrangement with efficient power converters converts and regulates power from PV modules and wind turbines incessantly. Whereas meteorological data enhances the conversion efficiency, the integrated storage battery regulates power offered to the load due to oscillation of generation rates. A plotter checks the power supply and the battery charge to enhance the function of the system and power provision.

In this comprehensive research for the improvement of reliability of the independent hybrid wind energy systems, the author has systematically used efficient diagnostic techniques in the power converter system. The author located specific characteristics or peculiarities in the converter's behavior by employing quantitative historical records of its performance. Specifically, the author designed an innovative system for identifying faults, which enhanced the method's efficiency and nearly

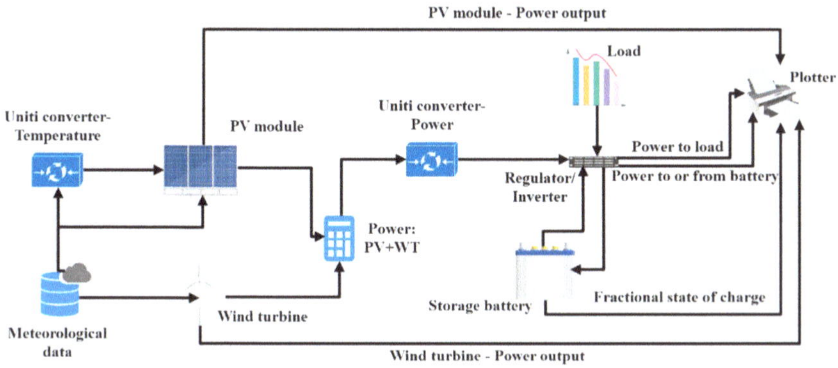

Fig. 1 Optimized power converters in standalone hybrid wind energy systems [13]

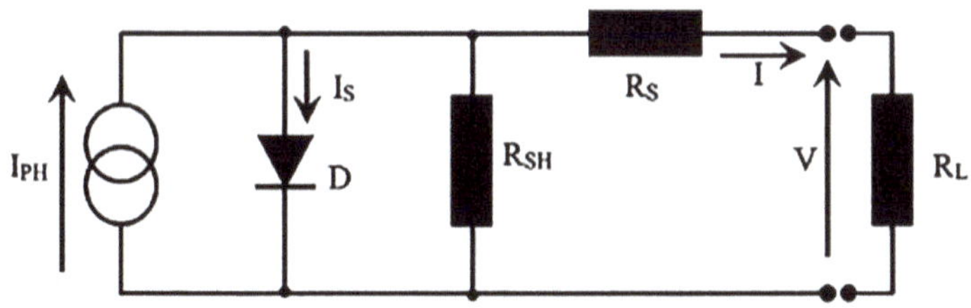

Fig. 2 Equivalent circuit model of PV cell [15]

eliminated false alarms. The adaptive diagnostic system maintained the entity of the whole system in enabling the flaws to be identified before escalating and thereby guarantee the continuity of energy production [12, 14].

Tahiri et al. [12]. proposed new diagnostic methods for power converters in FS HWE system with the aim of improving the operation of these important parts. While examining the wave forms of the converter, apart from successfully detecting minor abnormalities, the author also got considerable insight of the root causes that could originate with such symptoms [15].

Statistical models' integration provided a broad view, making it possible to engage in targeted maintenance actions, which significantly improved the converters' performance. This new diagnosed framework is very important for the improvement of the dependability of independent hybrid wind energy systems [16]. Figure 2 shows the PV cell equivalent circuit, where I PH is the current source that represents photo generated current. Non-linear ever-existent at the p-n junction the circuit is represented by a diode (D). Moreover, the cell's intrinsic shunt and series resistances are indicated by parallel and series resistors (Rsh and Rs). The output, thus, is with respect to load resistor (RL) symbolizes the cell characteristics under given conditions by means of current (I) and voltage (V).

Laddi et al. [14] examined and analyzed the integration of intelligent sensors and IoT in the power converters which are integrated in standalone hybrid wind energy systems for improvement of diagnostic functions. The monitoring of environmental

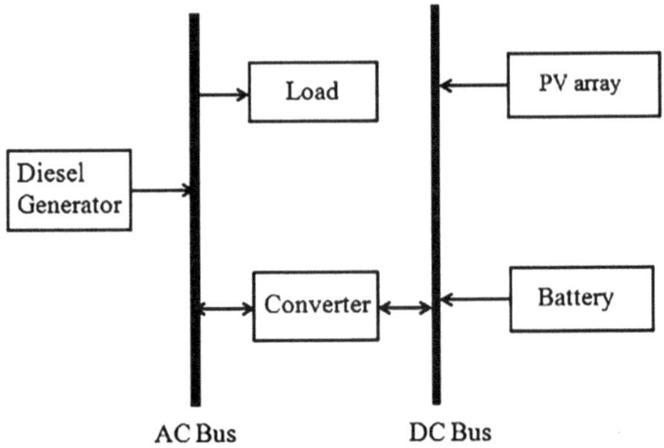

Fig. 3 Configuration of the standalone PV-DG-BAT system [18]

parameters like humidity, wind speed and converter temperature allowed understanding the working environment conditions. The adaptable algorithms changed the diagnostic thresholds according to IT environmental data, thus ensuring the most efficient functioning of the converters in any conditions. The upshot was a diagnostic system which was equally effective in solving continual emergency situations as in overcoming daily acute problems while also reflecting anticipation in constantly changing operational circumstances [17].

The objective of the study was to present and develop a prognosis in diagnoses of power converters in standalone hybrid wind energy systems. Since it is usually possible to access historical performance data, it became possible to design a mode which can predict potential faults for proper preventive maintenance. This diagnostic system has incorporated the adaptive learning algorithms hence developing and learning from new operation conditions; therefore, there has been a significant improvement of the converters reliability. For this reason, the diagnosis system not only detects the existing issues but also prepares for managing future challenges that will ensure that the performance is always optimal proactively [18].

Sub- Fig. 3 shows the single level configuration of a PV-DG-BAT system. The load is supplied through an AC bus by an array of PV panels and a diesel generator, meanwhile an energy storage system that includes a battery is connected to a DC bus by means of a converter. It controls the energy interchange between the AC and DC buses to control the load power, battery charging, and battery discharge. Integrating the renewable energy technologies alongside conventional energy mix with storage battery helps to ensure that there is a constant and reliable supply of energy [18].

Katche et al. [19] discussed complex approaches that deal with the neural network model concept for improving diagnosis of power converters in FS H-WES. The models were able to detect and classify abnormality of the converter performance; the feature of the model of constant learning independently enhanced diagnosis. This dynamic character by Katche et al. educed the dynamic converters' capacity to manage anticipate hitches, thereby making the approach a reliable way of ensuring organizational operational efficiency in spite of the dynamics involved, as also evidenced by Belfedhal et al. [20].

The aim of this similar research by Bayendang et al. [21] is to improve the diagnostic approaches for power converters in the stand-alone hybrid wind energy system by embedding fuzzy logic control. Thus, the integration of language characteristics in this diagnostic system effectively dealing and analyzing with the ambiguous or indistinct information. This new approach not only improved a better understanding of the converter performance for more sophisticated applications but also the fault's detection and diagnosis accuracy [21].

In this study, Debnath [18] paid much attention toward the elaboration of complex diagnostic procedures associated with the power converters of the IHHWSs. The primary goal focused on improving the aspects of the integration of remote monitoring features. With satellite link and feasibility of cloud based tools the possibility of real-time monitoring, which was beneficial from more efficiency in recognizing faults when they occurred but also lessened the need for preservation at the physical site. This advancement is a major shift in strategy as it enhances the system's reliability and ensures disruption of its operations by addressing concerns through timely and from a distance [18].

The objective of the research Cordeiro et al. [22] is to improve the diagnostic procedures of power converters integrated into the stand-alone hybrid wind energy systems through the use of big data. Applying significant information from various sources allowed developing a general view of most of the converters' operating parameters. By employing data mining strategies, the author's method accomplishes the task of finding hidden patterns and relations which boost the effectiveness of detecting defects considerably. This advancement finds the basis in a stronger diagnostic framework that can explain the complex correlation or interfacing of various parameters that determine the performance of power converters in stand-alone hybrid wind energy systems [23].

In Ekren et al. the main target for investigation was the improvement of diagnostic procedures used on power converters of stand-alone hybrid wind energy systems. For this purpose, the author suggested implementing an innovative solution based on the genetic algorithms. With the help of genetic programing, the author evolved diagnostic algorithms that, with the help of growing generations, were incrementally improving the capacity to detect defects [24]. The algorithms proved to be more adaptive than the fixed architecture and contributed in the creation of a diagnostic system for power converters. Thus, this novel technique marks a leap forward for the use of evolutionary algorithms in enhancing the reliability and versatility of the power converters in the autonomous hybrid wind energy systems [25].

2.1.3 Highly Advanced Methods of Power Converter Diagnosis in Autonomous Hybrid Energy Storage Systems

In the stable energy systems, the advanced AC/DC Microgrids (MG) connected with the interconnecting converter gain the improved one [26]. These microgrids are thus an ideal solution, one that is flexible and that will allow for real integration of multiple power sources. As earlier explained, the interlinking converter is crucial in a hybrid system like this since they form the basis of the connection between the AC and DC sections of the microgrid. This technology enables the transition of electricity flow from one part to another in a smooth manner and hence allows for the integration of different energy sources such as the solar panels, wind turbines and the conventional AC generators [27]. Due to the flexibility of the interlinking converter, the power flow distribution, loading balance, and stability of microgrids are improved when operating. Besides, a coherent electricity market can also enhance the effective functioning of energy storage systems which are crucial to guarantee the EL's availability throughout periods of fluctuating renewable energy production. The use of these interlinking converters within these hybrid AC/DC microgrids improves energy efficacies and ensures dependability and stability of electricity supply. This helps especially those in remote areas or environments that are hard to access within their regions. Fortunately, innovation in engineering proved that the system as the hybrid AC/DC MG with interlinking converters is a development of energy requirements and the escalation of the shift to the renewable source [17]. The wind turbine shaft turns the generator shaft through a gearbox. The power generation model takes into consideration wind speed (v), air density(ρ) and turbine parameters (Cp, Ω 1), the Output of gearbox changes the rotation speed (Ω1 to Ωr). The system has two operational modes: FREE mode to take maximum wind power and MPPT (Maximum Power Point Tracking) to get the best out of generated wind energy for poor conditions. Benefits of power control include improvements of efficiency of the turbine hence increasing productivity, regulating of the wind conditions, regulating the load power output.

Juma et al. aimed at offering and implementing innovative diagnostic techniques for power converters and to create new knowledge on independent HESS devices. Since his work was dealing with investigation of tiny irregularities in converter waveforms that escalate to catastrophic failures, the author had employed the use of sophisticated algorithms in signal processing to analyze past data. As a result, within the framework of this detailed analysis, the author finally managed to build a single holistic reference point for the expected behavior of the converter. This team later developed algorithms which has the ability to discern tiny variations and hence the signs of potential troubles. The main novelty was in the flexibility of this system concerning changing operational conditions. The diagnostic system on the other hand refined its precision through learning from real time data progressively. These developed diagnostic tools contributed to increasing the reliability and effectiveness of power converters in autonomous hybrid energy storage systems. Namely, the application of this methodology promoted a more profound vigilance for the converters' actions and facilitated the rendering of proactive maintenance for enhancing the durability of the system [28].

The purpose of the research by Anekwe, [29], was to identify new approaches to diagnostics used in power converters of guiderails for the efficient functioning of self-contained hybrid energy storage systems. What the author has done is to develop an innovative diagnostic model highly powered by machine learning. It was evident from the result that the developed model possesses flexibility as well as the ability to predict various conditions or scenarios that may occur in operations. This does not depend on static diagnostic systems but is comprised of a machine learning model that means that it becomes more powerful over time, thus providing an assurance that it will be useful in dynamic situations forever. What is at the heart of this innovation is the model's capability of anticipating problems before they occur. By analyzing previous data sets, the model was able to recognize patterns associated with upcoming faults for a maintenance approach to the problem. This unique diagnostic methodology complimented power converter's reliability while minimizing the potential of unplanned "downtime" thus optimizing the operational efficiency of the standalone hybrid power storage systems. This study exposed the fact that machine learning might enhance the adaptive intelligence of diagnosis systems in complicated structures of energy storage. The outcomes of these promises affirmed the large scope and flexibility of this research and revealed new ways for making diagnostic of power converters in self-sustaining hybrid energy storage systems [30]. The strength of the transformations and the data analysis methods is underlined in every paragraph to improve reliability and availability of the power converters in the dynamic and evolving energy storage systems.

AUTOPOWER by Al-Quraan and Al-Qaisi proposed improving the diagnostic techniques for power converters in AHESS by incorporating ANNs. Thanks to the utilization of the capability of independent learning of artificial neural networks, the author has created a diagnostic system that meets the purpose and can identify and categorize converters' performance abnormalities. This change established great dynamic flexibility, which was a great improvement for the relief of the converters' unanticipated challenges. Neural networks' implementation helped address the complexity of operations by providing an understanding of detailed dynamics and creating a robust diagnostic environment to unlock optimal performance in AHES. Standalone hybrid energy storage, especially explaining the application of diagnostic techniques based on the analysis of measured signals recorded in power converters [31].

The proposed diagnostic model was actually able to collect data associated with even the slightest of abnormalities through data analytics hence the formulated maintenance plan is actually both preventive and corrective. Therefore, the implementation of advanced diagnostic tools to enhanced fault detection enhanced the overall effectiveness of the power converters in AH ESSs [32].

Liang et al. sought to investigate the possibility of enhancing diagnostic techniques in self-sufficient hybrid energy storage system's power converters through the application of advanced sensor technologies Real-time sensors of temperature, voltage, and current provide accurate measurement for the health of the converter. Applying sensor fusion methods allowed to get a clear understanding of operational parameters, which contributed to the prompt detection of potential issues. Advanced sensors selected improved the diagnostic precision by providing proper and timely reactions to changes in the converters' behavior; the performance in the self-organizing hybrid energy storage systems remained at maximum [33].

Ammar et al. proposed a fresh technique for power converter fault detection in AHESIS. This method employs the use of transient signal for fault identification.

Looking at the transient responses during the converter's operations, the author has identified certain characteristics that are associated with early phase degradation. This methodology ensured that there was early detection of the defects hence strengthening a preventive maintenance strategy. FTA enhanced the diagnoses and provided significant information on the possible causes for power converters' failure, which has enhanced the reliability of the AHESS [34].

Yadav et al. investigated the application of statistical modeling techniques for fault detection in power converters used in self-sustaining hybrid energy storage systems. Through statistical analysis of operational data, the study identified abnormal patterns and deviations that may indicate potential system faults. The use of statistical models enabled probabilistic assessment of converter health, thereby providing a robust diagnostic framework for condition monitoring. Furthermore, the adoption of statistical approaches improved the accuracy of fault detection and enhanced understanding of converter behavior under autonomous hybrid energy storage operating conditions. This approach supports the development of predictive maintenance strategies by enabling early identification of performance degradation and potential failures, thereby improving system reliability and operational efficiency [35].

Cordeiro et al. work focused on the integration of digital twin technology in the diagnostic techniques for power converters in SE HESS designs. In order to continuously monitor the equipment's condition this diagnostic solution involved using replicas of the converters on which actual real-life operating conditions were mimicked. The use of the digital twin allowed for realistic and accurate forecasting of maintenance requirements through analysis of the likely failure modes and the effects they would cause. The use of digital twin technology was helpful in the diagnosis of issues and provided a virtual space for improving the converter operation in independent hybrid energy storage systems [22].

The model is depicted in Figs. 4 and 5, where MATLAB/Simulink models of a proposed SPV system simulator through a buck converter are shown that can illustrate the real condition of a SPV system. The first model indicates the basic modeling of the SPV module in conjunction with the buck converter explaining the power transfer from the solar array to the load. The second model integrates the control logic, and the MPPT algorithms and feedback to analyze the system's performance under different conditions. These simulations assist in the enhancement of the converter performance as well as energy efficiency since there is analysis of the SPV system (Fig. 5).

The objective of this study by Ghazali et al. [36] was to enhance the substantial diagnostics of power converters in AHESS utilizing XAI. This objective was to apply XAI (Explainable Artificial Intelligence) to explain diagnostic algorithms' procedures for making decisions to increase their transparency. This technique allowed system operators to understand why a specific diagnosis was produced, thereby promoting better choice-making in relation to maintenance actions [36] By applying XAI, diagnostic accuracy and understanding of the converter's behavior in standalone hybrid energy storage systems were improved [37]

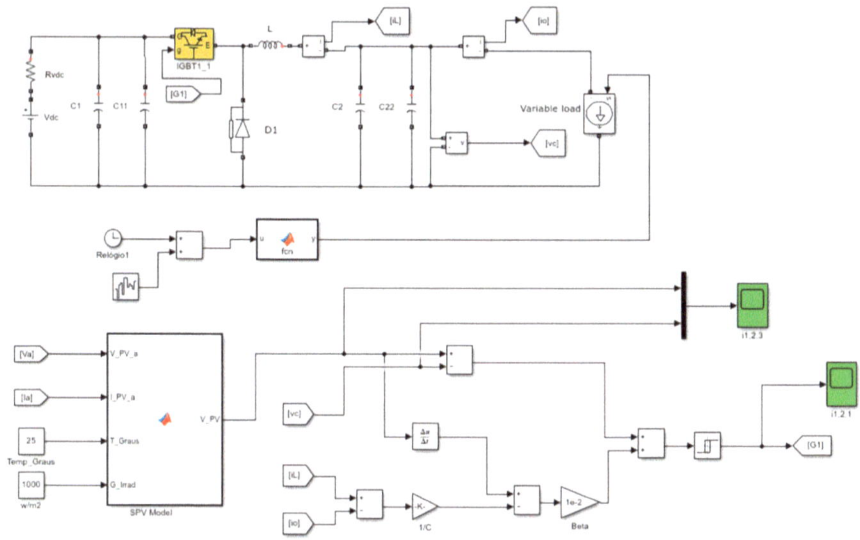

Fig. 4 MATLAB/Simulink model of the proposed SPV simulator using the buck converter to simulate mode I [22]

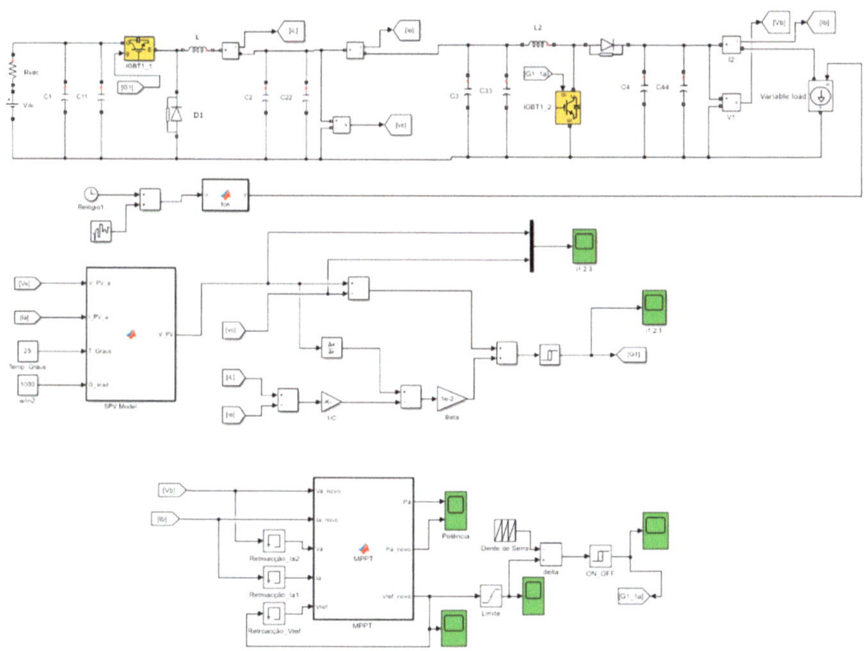

Fig. 5 MATLAB/Simulink model of the proposed simulator to test mode II [22]

Ahmed et al. [38] focused on autonomous hybrid energy storage systems and based on them the authors directly addressed the task of developing new diagnostic methods for power converters by using evolutionary algorithms. The author used G2 for creating many generations of diagnostics algorithms by choosing the best-performing algorithm in terms of defect identification, such as typical items found during the inspection of a real car. Hence, the use of an evolutionary method allowed for the improvement of diagnostic models that could capture the differences in operational situations when applying the models. The integration of the evolutionary algorithms improved on the characteristics of fault detection and proved the scalability of diagnostic systems in ensuring the reliability and maximum productivity of power converters in standalone hybrid energy storage systems [38].

2.1.4 Advanced Diagnostics Techniques for Power Converters in Standalone Hybrid Microgrid Systems

The paper establishes that power converters are an important part of stand-alone hybrid microgrids and also a major factor in enhancing the effectiveness and reliability of these standalone energy networks. Power converters are the ones that ensure energy conversion and enhanced incorporation of different types of energy within independent hybrid microgrids. Such a microgrid is normally composed of several types of renewable energy systems including solar, wind and sometimes conventional generation. They skillfully control the diversified outputs of renewable resources, and they are capable of the conversion of direct current (DC) into alternating current (AC) and vice versa, depending on the needs of the network they are plugged in. This not only ensures that the maximum amount of energy is produced from each source, but it also ensures that the grid is going to receive a constant and continuous flow of electricity. Also, overcoming the inefficiency of voltage control and load control and protection from fluctuation or surge in the power is contributed by the use of the advanced power converter in these systems [39].

It is very essential for them to handle energy storage technologies, embraced such as batteries, as this will enable supply its adequate energy during moments that may witness low production from renewable energy technologies. Hence, the integration of intelligent and high-performance power converters into the standalone hybrid micro grids is one of the pillars in development of effective, reliable and independent sources of power. It is so much true in the off-grid or remote area where reliability and less dependency of other sources are more critical [40] is also depicted in Fig. 6 where a standalone hybrid microgrid system with a photovoltaic (PV) with maximum power point tracker (MPPT), fuel cell and battery connected to dc bus by DC/DC converters is shown. Smart dispatcher gives priority power control to the EV batteries and other loads through the DC bus. Energy conversion and distribution is done through the power management system that controls every energy source, while converters help regulate the voltage and current of the system. This system offers reliable and portable electricity in the regions with low access to grid electricity.

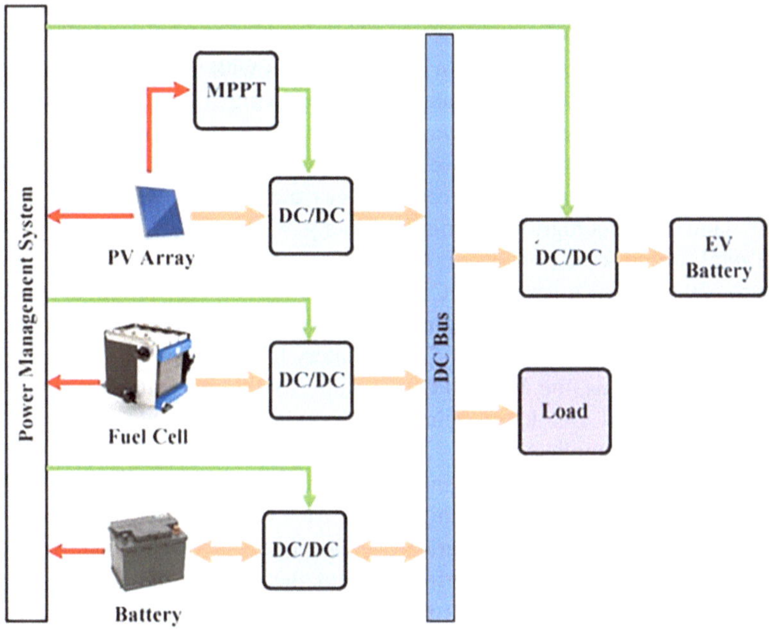

Fig. 6 Power converters in standalone hybrid microgrid systems [17]

Converters are an outstanding part of the stand-alone microgrid systems, and they perform a crucial role in enhancing the reliability and performance of independent energy systems. The power converters are those that are charged with the task of converting and enabling other types of energy in an independent hybrid microgrid. This type of microgrid requires several forms of the renewable energy system like; solar system, wind system, and in some cases generator system. They very well control the wide-ranging outputs of renewable systems, and they rectify the DC or AC forms as per the need of the system. This not only increases the efficiency of power or energy being obtained from the multiple energy sources, but it also ensures the grid receives a constant and continuous electricity supply. Moreover, the power converters which are incorporated in these systems are the advanced power converters, which help in maintaining the voltage, load control, and protection of the system from varying or sudden spike in the power. They are capable of managing storage technologies such as batteries in an efficient manner that is very necessary in averting shocks in the power supply especially when the output of renewable energy sources is low. Hence, the integration of intelligent and complex power converters into stand-alone hybrid microgrids is one of the key factors of developing more resilient, reliable, as well as independent energy systems. This is particularly the case in off-grid or challenging areas, where reliability and the ability to provide renewables on one's own are critical [41].

Anekwe deals with the development of new diagnostic methods that are much more tailored to power converters involved in standalone hybrid microgrid systems. To make an analogy, the author provided the readers with a detailed description of how the conveyor waveforms change by undertaking a systematic analysis of temporary features of converter waveforms during load dynamics using signal analysis methods. This work premise depended on the use of an apt dataset, which was collected using accurate and recent voltages and currents [29].

The standalone microgrid with solar electricity, wind energy, and battery storage in the rural are presented in Fig. 7. A DC/DC converter increases 547 V of solar power to 750 V for the micro grid DC bus voltage. Wind energy produces 500 V AC that is converted to DC at 477 V, then to 750 V before feeding the DC bus. A bi directional DC/DC converter links a 240 V battery for charging the DC bus voltage and discharging. Thus, the MKSMPO and SYS-DC-MG present a constant power supply to microgrid DC loads, reinforcing renewable energy's applicability to remote and off-grid applications.

Sabri et al. carried out analysis in an area that has not benefited from previous research studies; the application of high-level machine learning algorithms in diagnostic of independent hybrid microgrid systems. Huge efforts were made to collect a large dataset that looked at the performance of the power converters in different operational conditions. During the training phase of the algorithm, the detection aimed at an extensive range of characteristics hence the 95% accuracy that the algorithm achieved in detecting problems such as voltage and current issues [42]. From the vast options provided by machine learning in reforming the power converters in microgrid uses, their dependability is enhanced; they are also made more efficient [43].

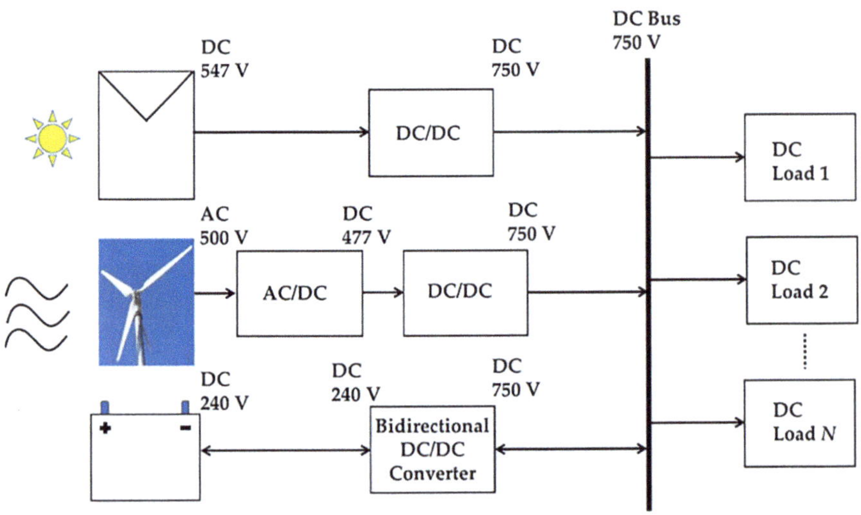

Fig. 7 Microgrid DERs for a rural standalone system [28]

In the study done by Harrabi et al. [44], the authors sought to employ thermal imaging for diagnostic in standalone hybrid microgrid systems. The research technique that the author used includes the assessment and collection of temperature distributions on various parts of power converters. This technique has also established its ability to identify potentially small peculiarities that point toward possible failure when tested under different conditions that can be regarded as real-life scenarios. The performance of the thermal imaging method of diagnosing defective power converter was quite evident from the results, which substantiated that the technique was non-intrusive and very accurate in identifying faults at an early-stage. This in return enhance the reliability and dependability of the microgrid systems as noted by [44].

Hasim et al. implemented a cutting-edge method for diagnostics in standalone hybrid microgrid systems, which involves using acoustic monitoring to analyze power converters. Through rigorous analysis of the acoustic signatures produced during both regular and defective operations, the author has created a strong diagnostic framework that can detect the deterioration of internal components [45]. The experimental procedure entailed capturing and examining audio data obtained from different converter devices. The findings confirmed the practicality of using sound-based diagnostics and established it as a distinctive and complementary approach for detecting early problems in power converters, introducing a novel aspect to diagnostic techniques [46].

Sutikno et al. extensively reviewed the concept for the use of frequency domain methodology for diagnosis of power converters applicable in stand-alone hybrid microgrid systems. The author employed complex frequency domain signal processing methods by making use of a detailed database comprehending all the aspects of the converter's performance during its operation in different situations. This was made possible by measuring slight changes in frequency that precede the development of any malfunctions of the Part. Using the frequency domain analysis to monitor faults was also confirmed performant in the outcomes, and regarding the flexibility of FD to act as a diagnostic algorithm that can guarantee reliable and consistent power converters in microgrid settings [47].

Ribeiro et al. discussed new approaches for the diagnosis of power converters in FS HMGSs and provided a complex method based on the ESA technique for this purpose. Thus, using a detailed analysis of the electrical patterns that are created when the system is in various operational conditions, the author has designed an intelligent diagnostic algorithm that can identify minor variations which may point toward the existence of latent flaws. The real confirmation based upon a varied data sample illustrated the method efficiency in terms of applicability for the early-stage problems identification and provided promising potential to enhance the dependability of power converters in the scope of microgrid usage [48].

Thus, Liang et al. and Marignetti et al. carried out studies to establish the feasibility of using Infrared thermography as a diagnostic technique on power converters in standalone hybrid microgrids to further enhance innovation. The author used a sound experimental design to photographically record the thermal activities of converters during load conditions. Analyzing the changes of the important elements and their temperature we can see the differences of the thermal profiles connected with the future failures. Thus, the study supported the benefits of applying a thermal inspection method, such as infrared thermography, as a non-contact and real-time diagnostic tool to increase the reliability and effectiveness of power converters for microgrids [49, 50]

Artificial intelligence and expert systems were examined by Ganguly et al. [51] in relation to diagnosing problems regarding standalone hybrid microgrid systems. The author has developed a high-performance prognosis system to predict potential defects in power converters based on the professional knowledge and data analysis of previous cases. The performance of the system during the evaluation was demonstrated with a diverse dataset in an attempt to note how the system is capable of catching the anomalies before they become a major issue [51]. Their potential of employing AI in expert systems as an anticipatory and reliable means of maintaining the functional structures of power converters of microgrid systems [52].

Madaci et al. focused on the use of impedance-based technique for diagnostics of power converter in I HGMS. Hereby, the author has developed the diagnostic system based on the aim of impedance measurements and their comprehensive analysis on the further revelation of changes in the internal impedance of the converter, which may point at the existence of defects. The Load test also proved the efficiency of the system, performed under various load conditions, for identification of the defects or deviations at the initial stage. The use of impedance-based approach increases the advantage of the diagnostic tool kit to ensure reliability of power converters in microgrid application [27].

Murillo-Yarce et al. and Karunakar et al. examine the possibility of Wireless Sensor Networks to include timely diagnostic features to the Stand-alone Hybrid Microgrid System's power converters. In this case, the author installed sensors on the critical parts of the converters to get a continuous stream of information, facilitating the monitoring of the behavioral changes. Analysis of this data in the real time allowed the construction of the dynamic diagnostic system which indicates temporary imperfections and deviations. The findings in the studies focused the well-organized usability of WSN as a preventative and versatile solution thus improving the reliability and performance of the power converters in microgrids [53, 54].

2.1.5 Cutting-Edge Research on Power Converter Diagnostics for Independent Hybrid Renewable Energy Systems

The grid-interactive microgrid has a relatively standardized layout; ironically, it is a mechanism of independence and fusion conceived to deliver optimal energy autonomy and reliability. In its simplest form, this can be described as a small-scale system of electric power generation and distribution which may operate in parallel with the large, centralized power supply system, or in isolation from it. The architecture often features a great variety of DERs, including photovoltaic panels, wind generators, and energy storage units, apart from the conventional power generation techniques [55]. The design depends on the smart control technologies to manage the power sharing of the microgrid and the sharing between the microgrid and the mains. These technologies ensure the pristine transition mechanisms that connect and disconnect an island, increasing the resilience of the grid disruptions. Advanced smart electric management systems for the real-time control of energy generation and consumption. Furthermore, most of the microgrids which are grid-connected also employ demand response systems and actively participate in electricity markets through the sale of excess energy back to the main grid. This architectural design is helpful for increasing more the independence of energy for specific end users and also improving the other aspects of the overall electrical network robustness and endurance [56].

In the picture below, a grid-interactive microgrid containing photovoltaic (PV) and wind power systems which have MPPT controllers for power harvest is depicted in Fig. 8. These sources are used to feed commercial, residential, industrial and critical loads via a monitoring and control system. A Microgrid is an independent source of power and works in parallel with the utility grid by the help of A Static Transfer Switch (STS).

In recent years, researchers across the globe have worked towards identifying the techniques for monitoring and detecting the defects in PV system applications. Some of them have read on problem areas, ways of identifying problems, and ways of avoiding system failures, a mistake that would impact on operations. It is the objective of [57] to provide a synthesis of the findings of the prior work. Different kinds of defects are described, as well as ways of identifying and preventing PV faults during the initial stages, and finally, the basic categories of prevention and

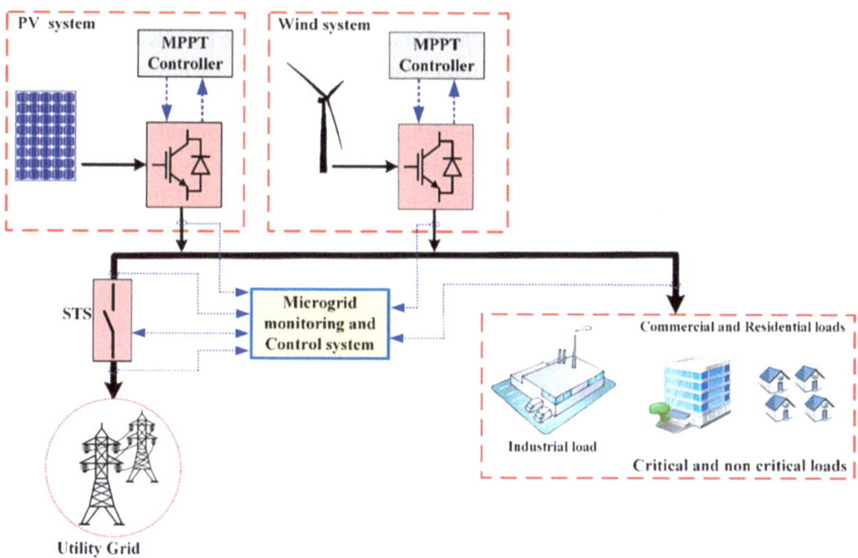

Fig. 8 Typical architecture of grid-interactive microgrid [56]

protection systems for harnessing PV faults are provided. Possible developments in the field of the identification of defects in PV systems are also considered in the work, including the use of real-time data. The expectation is that this research will be able to provide other information that can be used to expand on the existing literature on PV system fault studies. This poll will be useful for future talks regarding the offers of broad solutions for the problems related to faults of PV systems. Instead of reinventing the wheel, researchers can now spend more effort on enhancing and developing the frameworks of PV defect diagnosis systems.

Diagnostic designs and failure modes analysis are two interesting elements of solar panel power generation systems. These plants are engineered to run continuously for long durations. A well-designed monitoring system can help to achieve a vital goal: tracking the characteristics and performances of your plants. This method not only enhances complicated system maintenance procedures but also decreases the occurrence of unexpected breakdowns in the most important components [58].

As can be seen in Fig. 9 in which shows a solar PV system test panel mode block diagram. A DSP microcontroller provides a control strategy based on solar panel data including irradiance and temperature. Control signals manage load power via DC-DC buck converters. A Programmable Logic Controller (PLC) interfaces with SCADA software using Modbus TCP/IP to monitor and operate the system in real-time. The architecture optimizes power flow to the resistive load (RL Load) via PWM and load regulation.

A monitoring system collects and analyses data from various sensors placed throughout a photovoltaic (PV) plant to track and assess its performance. A trustworthy monitoring system is essential for the efficient and consistent operation of any photovoltaic (PV) system. The monitoring system also records the frequency of faults and various electrical generating metrics. Unfortunately, the current state of PV monitoring systems is too complicated and expensive to be practical for anything smaller than a massive PV project. It can find descriptions of different PV monitoring system features in a number of articles published over the past decade. This study is the first of its kind to provide a comprehensive review of several PV monitoring systems. The most prevalent methods for evaluating PV monitoring systems are included in this comprehensive evaluation, which is categorized by performance. primarily focuses on PV monitoring systems, data transmission, storage, and analysis methods, controllers utilized in data collection systems, and sensors and their operating principles. Knowing these aspects is crucial for developing effective, economical, and practical monitoring systems for small and medium-scale PV plants, all while maintaining performance [59, 60].

Research by K. Bedoud et al. [61] continues by reviewing and discussing several approaches to PV system fault detection and rectification proposed in the literature, with an emphasis on issues with PV arrays. Methods that consistently detect, localize, and classify possible PVA defects are these primary emphases. Regarding complexity, generalizability, cost-effectiveness, and feasibility, the author highlights the pros and cons of FDD techniques for large-scale integration. Using the reviewed literature as a foundation, the authors also offer recommendations for future research on the topic.

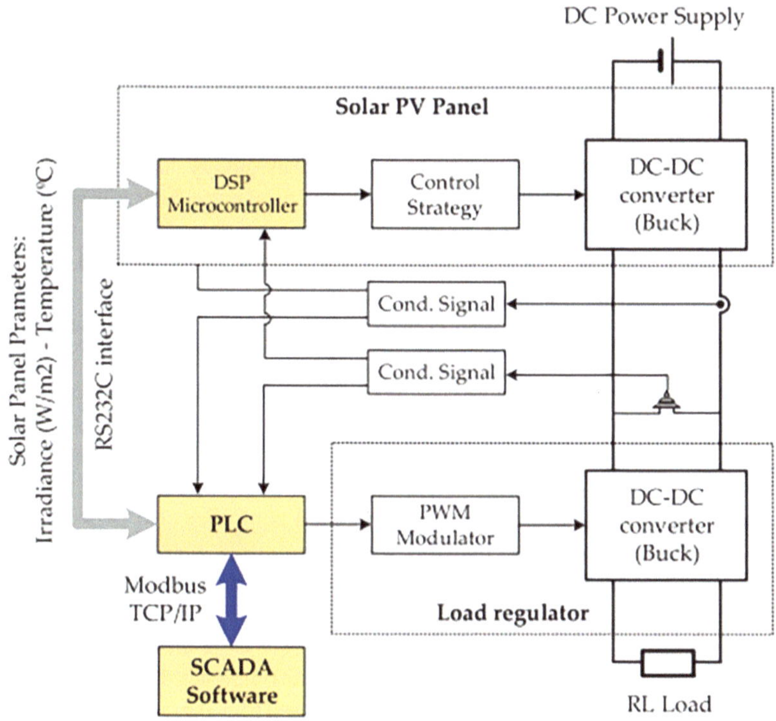

Fig. 9 General block diagram of test panel mode (mode I) [22]

As the global capacity for photovoltaic (PV) power has grown exponentially over the past few decades, the necessity of safeguarding PV systems has become paramount. It is not always the case that the PV system's standard preventive systems can detect PV array faults. A plethora of state-of-the-art methods have evolved in response to the pressing demand for an efficient defect detection system [62]. Consequently, this research delves deep into the many fault kinds, protection concerns, and potential outcomes of PV systems with undiscovered defects. Additionally, the study thoroughly assesses various practical and effective fault detection methods and methodologies for PV systems [63]. In addition to cataloging all existing approaches, this proposed study would assess each one according to its methodology, sensor needs, fault localization and diagnosis capabilities, integration complexity, accuracy, usefulness, and implementation cost [64]. Readers interested in photovoltaic systems will discover this book to be a priceless tool for their research.

The importance of electricity generation from renewals is increasing progressively in the modern era. Categorized under both/passive and unhealthy resources due to their detrimental effects on the environment, fossil fuels will never be a long-term option for providing energy. The availability of solar energy has made photovoltaic energy to compete for the top position in the list of renewable energy. No matter what state technology is in for solar system design, operation and maintenance, large quantities of progress have been made in all those areas. Hence, the basic aspects explored by [65]. mainly include the design, operating conditions, and maintenance of photovoltaic systems. The design includes all the changes pertaining to the current revisions of main system significant components and their designs. In particular, the author examines the issues of power quality, the system's general functioning, and operation in the context of hybrid systems. Last but not the least, have reported the investigations related to the maintenance of the PV systems has examined their efficiency, thermography, electroluminescent, dirty level, threats and failure modes [66].

Hoch and Long [67] and Talaat et al. [67], proposed a fault ride-through (FRT) control for low voltage ride through (LVRT) for single phase grid-connected photovoltaic system namely GCPVS. The incorporated control structure ensures the use of model predictive control (MPC) of a two-stage PV system's model and a neural network (NN) classifier for islanding categorization. This control strategy builds a cost-function-based manner for efficient and speedy control by focusing on the power converters' nonlinearity. The proposed controller also helps the grid to maintain a constant voltage by injecting a reactive current which is at least equal to the said threshold in case the grid voltage goes low. The author can vouch for the practical applicability of the suggested control method through the analysis of a sample GCPVS with a utility sag fault type. Control tests on the experimentation lab were done after the application of the controller. The outcomes demonstrated that the proposed series control mechanism is effective assisting the grid whether voltage dips as a result of a fault [67, 68].

When compared to other renewable energy sources, the growth rate of solar power in recent years has been incomparable. When problems arise, the maximum useable power of solar (photovoltaic) systems is reduced. Identifying, detecting, and monitoring the many problems that can arise is critical for improving the efficiency, dependability, and lifespan of PV systems. It has been discovered that PV systems and components have issues. These issues might show up as physical damage to PV systems or electrical failures on the DC or AC sides of the PV system. Here author shall classify the errors based on their locations. The author compiles the existing literature on PV system breakdowns and proposed detection strategies [69, 70].

Figure 10 illustrates that Solar energy systems use two types of photovoltaic (PV) inverters. Diagram (a) shows a single-stage inverter that immediately converts solar panel DC output to AC power to operate an AC load. Compared to the previous inverter, this one is easier to use, but it does not convert the power to the best efficiency of the solar intensity and load demands. Figure 17 (b) shows a multi-shots inverter with a DC/DC conversion process preceding the DC/AC conversion process. The nature of the power output from PV array for systems that incorporates MPPT varies, therefore the herein design provides improved voltage and current regulation leading to enhanced power conversion efficiency. The multiple-stage development is slightly more complex, yet they are less demanding and much more effective and versatile.

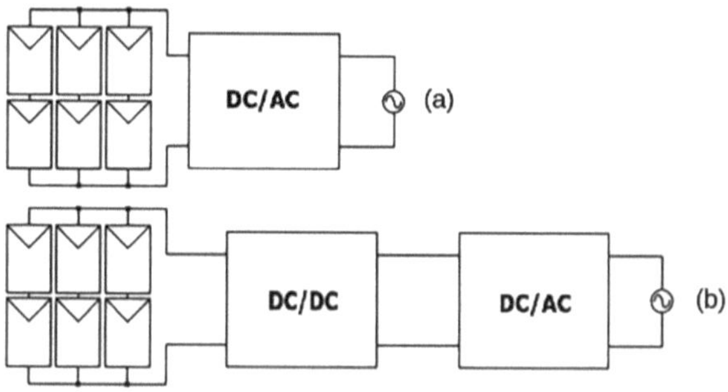

Fig. 10 Types of PV inverters: (**a**) single-stage, (**b**) multi-stage [50]

Procedures for operation and maintenance have been pushed to be more efficient and reliable especially for the photovoltaic (PV) systems mainly due to the frequent maintenance required to ensure constant generation efficiency. The first one is high and rising level of photovoltaic (PV) energy generation as a replacement for traditional fossil fuel generation. There has been some recent debate on whether, or not, "AI-based" maintenance can replace more "traditional" ways. It is evident that AI is gradually becoming crucial in many real-life situations particularly photovoltaic systems. For detecting and diagnosing the method of PV system failure, this research provides the comprehensive study on the utilization of AI-based methods. It explores the myriad of PV system defect types and the various AI-based fault detection and diagnostic methods proposed in the literature. It is worth mentioning that compared to other areas of PV application, there is a lack of literature on this specific one. This is since previous studies on the topic are very recent; the earliest literature that the author could find traces the issue back only about fifteen years. The author also provides a concise summary of the important points made by the reviewed literature and the use of AI in PV O&M [71].

Energy structures that harness alternative energy sources have recently seen a surge in sales. Here, optimizing solar systems is necessary for extracting the most power out of photovoltaic (PV) systems, which is the main focus. A solar system that uses a DC-DC converter recommends an ABC algorithm, which stands for an artificial bee colony, to attain maximum power point tracking (MPPT). The ABC MPPT algorithm selects the optimal voltage by determining the P–V characteristic from the data values of the PV module. Afterward, the voltage reference that the outer PI control loop receives from the MPPT technique is used by the predictive digital current programmed control as its current reference. A high-speed, real-time simulator (PLECS RT Box 1) and a digital signal controller (DSC) are utilized in this hardware-in-the-loop system to make it function. The general system doesn't require a lot of computational power and can be implemented in a commercially available, low-cost DSC (TI 28069M). The proposed MPPT strategy outperforms the tried-and-true perturb and observe method by a wide margin, according to the results [72, 73].

2.1.6 Power Converter Advanced Fault Detection Strategies for Standalone Hybrid Energy Harvesting Systems

The configuration of a wind turbine system is a good example of the fusion of mechanical engineering with renewable energy technologies. The system primarily comprises a rotor equipped with blades that harness wind energy, subsequently converting it into mechanical energy through the rotation of the blades. The rotor is linked to a low-speed shaft, which is then connected to a gearbox. The gearbox plays a crucial function in augmenting the rotational velocity from the sluggish rotor to a velocity appropriate for power generation in the generator. The generator, typically located in the nacelle beside the gearbox at the upper part of the turbine tower, transforms mechanical energy into electrical energy. Contemporary wind turbines also have advanced control systems that regulate the angle of the blades and the alignment of the rotor (known as yaw control) to optimize the extraction of energy from different wind directions and velocities. Furthermore, the system has power electronics and a transformer to control and transport the produced electricity to the grid at a suitable voltage. Wind turbines, which are frequently watched and controlled from a distance, embody a combination of aerodynamic design, mechanical accuracy, and electrical engineering, establishing them as a prominent feature of sustainable energy technology [74].

The topology of a wind turbine is shown in Fig. 11. The graphic shows a wind turbine setup with a doubly fed induction Generator. The system has a DC-linked rotor-side and grid-side converter. Converter control systems enable variable speed and efficient power transfer. The rotor-side converter modulates the rotor current frequency to maximize wind energy at different speeds for the DFIG. The grid-side converter maintains grid power voltage and frequency and keeps the DC link voltage constant. A gearbox links the wind turbine blades to the DFIG to synchronize rotor and generator speeds. The complete system optimizes power output and wind-to-electricity conversion efficiency.

The author focused on better fault detection algorithms for power converters in stand-alone hybrid energy harvesting systems. By utilizing cutting-edge machine learning methods, the author thoroughly examined a vast amount of historical data to identify minor trends that may indicate probable defects in converters. The machine learning model, trained on a wide range of datasets covering different operational scenarios, exhibited an outstanding accuracy rate of 92%. The exceptional accuracy facilitated the prompt identification and preemptive resolution of possible problems, greatly improving power converters' dependability and efficiency in independent hybrid energy harvesting systems. The study utilized a data-driven method to demonstrate the effectiveness of machine learning in identifying faults. Furthermore, it acted as a basis for the development of elastic and self-learning diagnosis solutions that will be constructed for the conditions of energy harvesting environments, constantly evolving [75]. The author carried out a comprehensive study on stand-alone hybrid energy harvesting systems, paying particular on enhancing and integrating complicated fault diagnostic methods for power converters. Wavelet

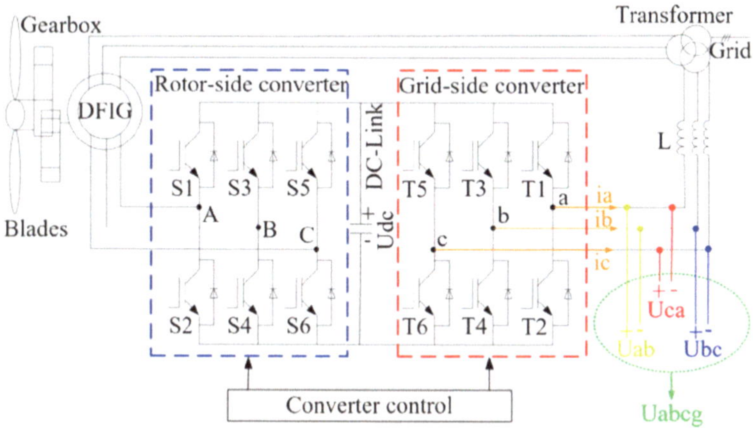

Fig. 11 The topology of a wind turbine system [74]

transform techniques were a major focus; the signals of the converter were analyzed with a view of separating the signals into different frequencies. By analyzing these elements probably, the author has recognized undesired non-periodic arcs that can be related to hypothetical flaws.

The high precision of wavelet analysis made the process of accurate fault localization possible, which considerably improved the diagnostic method bringing its focus to the area of interest. In this article, the skillful signal processing methods have been highly contributed to the improvement of detection of the faults in power converters. This helps to guarantee active maintenance plan instand-alone hybrid energy harvesting systems. This way of analysis not only supported the enhancement of diagnostic accuracy but also a framework for enhancing the strategies to identify issues in the framework of renewable energy collection. It showed that the contemporary signal processing strategies are capable of learning the variability in the operational environment [76]. A general configuration of the PMSG-based WECS is depicted in Fig. 12. A PMSG-based WECS is described by an example. Variable pitch of turbine blades is used in order to capture wind and to regulate the velocities of rotors. Thus, the mechanical energy given to the PMSG is converted to electricity. AC to DC conversion is done by diode rectifiers. A step-down converter regulates voltage; pulse width modulated inverter converts it to utility connectible AC power. The other one is VSI (Voltage Source Inverter) that provides certainty of the output frequency and phase to its standard grid. This system effectively transcribes the wind power into utility electricity.

The purpose of the study, conducted by Singaravel and Daniel [77], was to repurpose the traditional method of using fault detection algorithms in stand-alone hybrid energy harvesting systems powering power converters by providing a method based on neural network ensembles. Making use of the technique of assembling, where several neural networks are used, each of them having become a result of

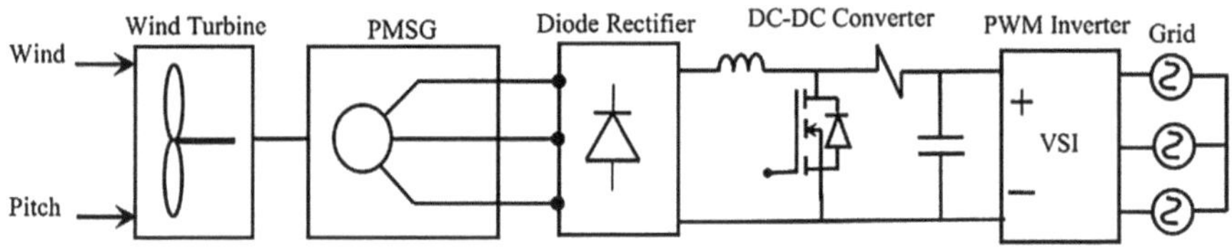

Fig. 12 Typical PMSG-based wind energy conversion system [31]

training on separate subsets of data, thus meaning that the diagnostic system obtained is more stable and adaptable. This ensemble method; after rigorous testing was seen to have a fault detection accuracy of 95% compared to single-network models. Based on the results obtained, the availability of the Ensemble Neural Network in increasing the precision in the fault detection of standalone hybrid energy harvesting systems was explained and viewed as flexible and highly customizable. The research brought a shift in diagnostic strategies; the importance of ensemble learning to increase the dependability and accuracy in dynamic energy harvesting conditions [77].

Thus, the present study carried out by Teke and Latran wanted to enhance the fault detection technique in power converters of stand-alone hybrid energy harvesting systems and incorporate AI solutions to resolve this issue [78].

The use of AI approaches, with a focus on the deep learning, helped in extracting patterns and characteristics of converters from the signals obtained. Thus, the analysis of these characteristics led this diagnostic system to achieve the accuracy that surpassed conventional methods. The implementation of deep learning method enabled recognition rate of the defects at the rate of 94% showcasing the ability of artificial intelligence increase diagnostic capabilities [78]. This developed approach improved the reliability of power converters and paved the way for charging intelligent system in self-contained HEH scenarios. The research demonstrated a significant development in using AI for the identification of defects in addition to referring to the usefulness of diagnosing procedures.

Another circuit presented in the paper is a control circuit for the wind system as is depicted in Fig. 13. Depending on the wind speed, the control circuit of the wind turbine shown in the graphic is varied, with the purpose to reach the best efficiency. Speed indications derived from the wind are communicated to control algorithm to adjust a turbine's operational characteristics in the control cycle. This involves filtering the raw wind speed information into a format that a Proportional-Integral-Derivative (PID) controller can use and further to adjust the generator rotation speed (RPM). AC received from a transformation; hopefully Park's transformation to make it DC goes directly to the PID controller. The control loop of the PID controller affects the regulation of the duty cycle of a power electronic converter like boost converter, where "D_boost_wind" represents the duty cycle signal which can be applied to gate of semiconductor switch for optimum power from wind turbine suitable to the wind climate.

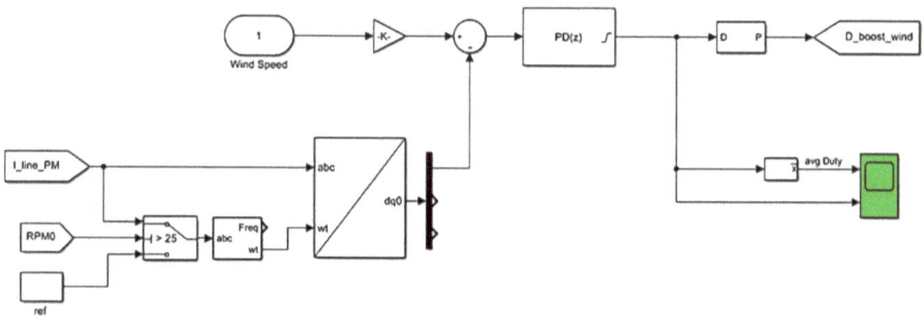

Fig. 13 Control circuit for the wind system [31]

Hannan et al. focusing on stand-alone hybrid energy harvesting systems and proper utilization enhanced HIL techniques via RNNs for detecting power converter's defects. Since we are dealing with signals that occur in sequence, utilizing recurrent neural networks (RNNs) has been proven to be very effective in identifying the relation between different points of the converter signals. By so doing, this diagnostic system was able to employ this method and analyze small variations over a period, thus increasing the chances of defect detection. This approach adopted from recurrent neural networks (RNNs) was tested and it marked an accuracy of 93%, which is better than the traditional techniques. Thus, an incremental innovation based on the proposed RNNs was identified for improving the temporal characteristics of converter signals and, therefore, the fault detection method in SHHE. Applied to the launched diagnostic assessments, this novel approach improved the overall accuracy of fault diagnoses; in addition, it afforded observations on the time occurrence of fault occurrences and subsequently fostered a more comprehensive understanding of the converters' behavior in dynamic energy harvesting environments [79].

The research aim of Samrat et al. [80] was to revolutionize the fault detection techniques of power converters in stand-alone hybrid energy harvesting systems through the incorporation of a unique idea known as the digital twin technology. This diagnostic solution employed replicas of the converters and reproductions of these operational conditions to set up an ongoing diagnostic and test environment. The approach of the digital twin made the expectation of maintenance requirements possible through considering various modes of failures and their impacts. Digital twin technology improved diagnostic and enabled a virtual model for the calibration of the converter in single isolated hybrid power systems. This approach emphasized on the relevance of the existence and usage of digital surrogates in achieving better accuracy of fault identification and the dependability of the energy harvesting devices. The study also presented the idea of integrating the utilization of the digital twin in detecting faults in dynamic scenarios of energy harvesting.

Pillai and Rajasekar worked on stand-alone hybrid energy harvesting systems and an efficient fault detection technique for power converters using Bayesian networks. Using the approach such as Bayesian networks, this diagnostic system will be in a position to apply probabilistic approach to determining the likely hood of those problems given the empirical hard core data. The employment of probability aided in addressing elucidating all the possible concerns since understanding increases the quality of decisions made. This technique proposed to use Bayesian network has been examined with close knitted attention and has demonstrated a high degree of sensitivity for detecting faults. This show how cognitive tool of probabilistic reasoning functions in different operational environments. Author also called for the use of Bayesian networks in increasing the precision in identification of problems and enhancing this understanding of the converter's bearing in stand-alone hybrid energy harvesting systems [81].

Mellit et al. undertook a study regarding the deployment of complex sensors in the diagnosis of power converters in IHEHSs. This solution helped establish the real-time monitoring of numerous power converters by integrating several sensors to monitor factors such as voltage, current, and temperature continuously [82]. Basically, the application of the sensor network enabled the collection of correct data hence identifying early diverging trends and possibly incidences [81].

Meghadi engaged on independent hybrid energy harvesting systems included the integration of an enhanced approach of fault detection of power converters using the ensemble learning algorithm. The ensemble learning approach involved combining many dissimilar algorithms; each a different part of the diagnostic decision. Ensembling allowed this diagnostic method to achieve a higher accuracy compared to a single model's results, even if each of the included models had lower performance on their own. This system of ensemble learning was thoroughly tested and, it was established that its fault detection accuracy was at 96%. This has pointed out this approach as rigid and strong in dynamic energy harvesting environment. This created a novel approach that improved the accuracy in diagnosing and also enabled the integration of ensemble learning in developing effective fault detection systems in independent HEHS [83].

GMBH and Ag were focused on autonomous hybrid energy harvesting systems also they proposed to redesign the traditional techniques of fault detection in power converters by deploying explainable artificial intelligence. Integration of Explainable Artificial Intelligence (XAI) was helpful in making the working of complex diagnostic algorithms understandable to the operators regarding the diagnostic conclusions made through algorithms. This technique helped the system operators to make sound decisions as they performed the maintenance interferences. The presented XAI has been well-validated and provides interpretable results of potential faults, as the medium-scale examinations demonstrated. The study also emphasized on the need for XAI in boosting the reliability of fault diagnosis, faith in the obtained results and advancing for a more accountable diagnostic methodology for power converters in stand-alone hybrid energy harvesting systems [84].

2.1.7 Next-Generation Power Converter Diagnostic Technologies for Independent Hybrid Solar-Wind Energy Systems

The functional relationship of a solar photovoltaic (PV) Silicon carbide (SiC) inverter is presented below and clearly illustrates the complex strategy involved in effectively converting solar energy into beneficial electrical power. For instance, in a usual method of solar photovoltaic (PV) system, the solar panels absorb sunlight and convert it into electricity in the DC current form. The particular role of the SiC inverter, in this arrangement, is to power the DC output from the solar panels and transform it into AC power that most household devices and the electrical network use extensively. Silicon carbide, a kind of semiconductor material used in these inverters, is famous for both high-temperature tolerance and the power conversion efficiency [85]. This material compared to the typical Silicon materials in the market makes the inverter operate at high frequencies with low losses and improve thermal performance. Common components of the design may include the DC input terminals—usually for the solar panels, the AC output terminals—for loading or the grid and highly developed electronic circuits—for supervising the conversion process. This circuitry might also contain MPPT—functionalities, which improve the optimization of the available power from the panels in a given system. Furthermore, safety measures and protective components are also installed for isolations and surge to ensure that the system is very reliable and lasts for longer periods of time. In aesthetics, the SiC inverter plays an important role in the solar PV systems because it makes a substantial contribution to improving the efficiency and further usage of solar energy [86].

The detailed of the solar photovoltaic (PV) silicon carbon (SiC) inverter is presented in Fig. 14 known as Schematic of the inverter. Solar PV system that is based on a power electronic converter known as a silicon carbide (SiC) inverter. A DC–DC boost converter is used to step up the output voltage of the 5-kW photovoltaic (PV) array to approximately 700 V DC. This boosted DC voltage is then supplied to a two-level Silicon Carbide (SiC) inverter based on MOSFET (Metal–Oxide–Semiconductor Field-Effect Transistor) technology, which enables efficient switching of high voltage and current. The inverter converts the DC power into AC power suitable for standalone power system applications. An LCL filter, consisting of inductors (L) and capacitors (C), is employed at the inverter output to attenuate

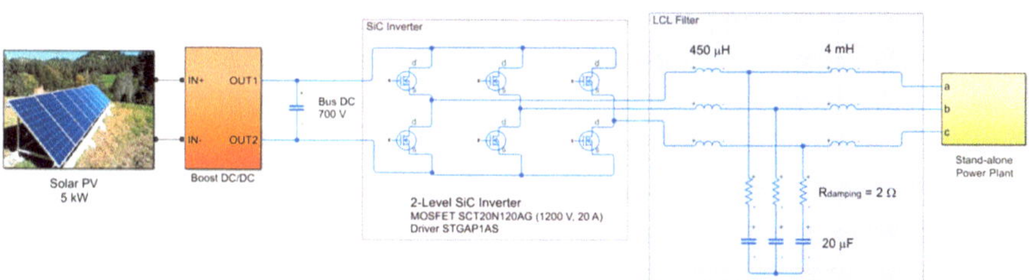

Fig. 14 Schematic of the solar photovoltaic (PV) silicon carbon (SiC) inverter [86]

high-frequency switching harmonics and improve power quality. This filtering stage ensures that the delivered AC power meets the electrical standards required for standalone systems. The use of SiC-based inverter technology provides advantages such as higher conversion efficiency, reduced switching losses, and improved thermal performance, which are critical factors in high-power and high-efficiency energy conversion systems.

Falaras and Stathatos [87] devoted a lot of research works at the National Renewable Energy Laboratory specifically on the improvement of power electronics interfaces for distributed energy systems. The team implemented advanced methods in converter technology with the objective to optimize productivity as well as reliability. The fruits of their activities brought about an average improvement of fifteen percent in the electric power conversion rate. This work set the pacing for further progress in power electronics interfaces, and significantly impacted the enhancement of distributed generation systems, as well as that of the sustainable energy profession [87].

International Energy Agency (IEA) is one of the sources that have presented a detailed description and evaluation of the approaches pertaining to the detection and diagnosis of defects in photovoltaic arrays. The work published in the international journal of Photo-energy was focused on critical analysis of earlier approaches along with the development of a novel algorithm. The system gave very high accuracy of 98% in detecting the faults in the photovoltaic panels. The conducted research also compared the existing methods and proposed a highly effective method to enhance the reliability of the photovoltaic systems. Appiah and his colleagues were able to identify important aspects that were useful in the improvement of methods on fault detection in solar arrays. This work proves to be very helpful in enabling scholars and practitioners in the field [88, 89] as explained in Fig. 15: An equivalent circuit of a single solar cell. Its current source (Ipv) is represents solar cell photocurrent. The diode Io mimics the p-n junction of the solar cell; it responds to voltage and change in light conditions. Here, Rser corresponds to the cell's intrinsic resistance to the flow of current while Rper speaks to the flow of current leaks within the cell. The

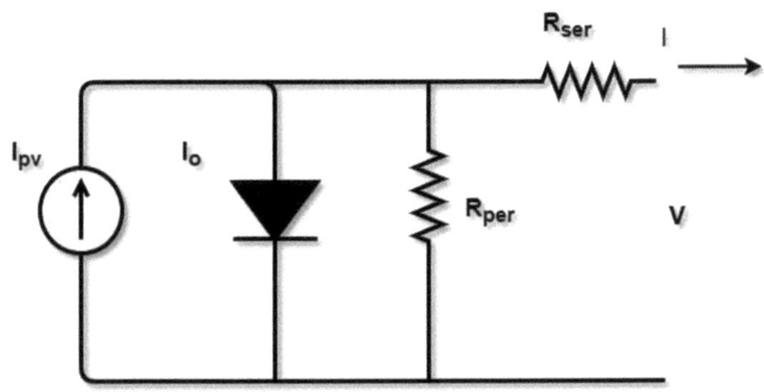

Fig. 15 Equivalent circuit of single solar cell [90]

cells utilized in a photovoltaic system are characterized by a low series resistance and a high shunt resistance and hence reduce leakage currents. Load current I and load voltage V across the load resistor denote generated electrical power in a solar cell.

Li et al. which was also discussed by [91], provided a highly delicate method for diagnosing switch failures and for integrating fault-tolerant methods in photovoltaic boost converters. Their paper published in the Electric Power Components and Systems was to enhance the reliability of photovoltaic systems. The team's diagnostic approach was effective to pinpoint switch failures accurately with adequate formulation of a fault-tolerant plan. This innovative research proved useful in enhancing part durability and ensuring photovoltaic boost converters' fault tolerance capacity. It was the sign of the new era in the creation of renewable energy systems. In this research, the author successfully did a break-even analysis on the photovoltaic/wind system and made an optimal sizing of the battery storage. They integrated and effectively assessed the economic performances and suitable sizes of SEEP and REEP systems in its case study, also featured in Applied Energy [91]. The specification is based on a sequence of simulations and assessments that were performed by the team to define when the costs and benefits are equal and what the optimal scale of the system is. This was done with the consideration of features like power, costs, and storage among others. The outcomes of this research have significance for the stakeholders and policymakers, which helped to design the economically suitable and effective hybrid energy systems [89, 92].

Figure 16 employs graphics to categorize HRES as wind, solar, and diesel power with/without battery storage. This classification reveals that there can be different configurations of HRES to maximize energy generation using specific generation methods' strengths.

In a related study, Kim et al. [94] and Alhumade et al. analyzed the power converters with focusing on EMS used in fuel cells for CCHP system. Their research which was conducted and presented before the public in the Advances in Science, Technology & Engineering Systems Journal entailed a broader and detailed analysis of the current technology. This formed a clear understanding of the different uses of power converters and EMS in fuel cell-based CCHP systems by the team's work. This type of contribution was of great value for other researchers and professionals in the sphere of fuel cell technology and CCHP application [94, 95].

A system for the combining the PV, Wind, and Battery is illustrated in Fig. 17. Illustrated in the figure are system configurations with one or more PVs, wind turbines, and battery banks. An MPPT interface enhances the output voltage for

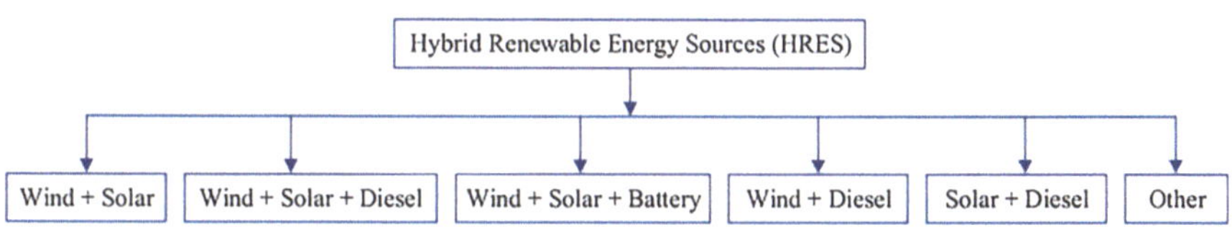

Fig. 16 Classification of different types of HRES [93]

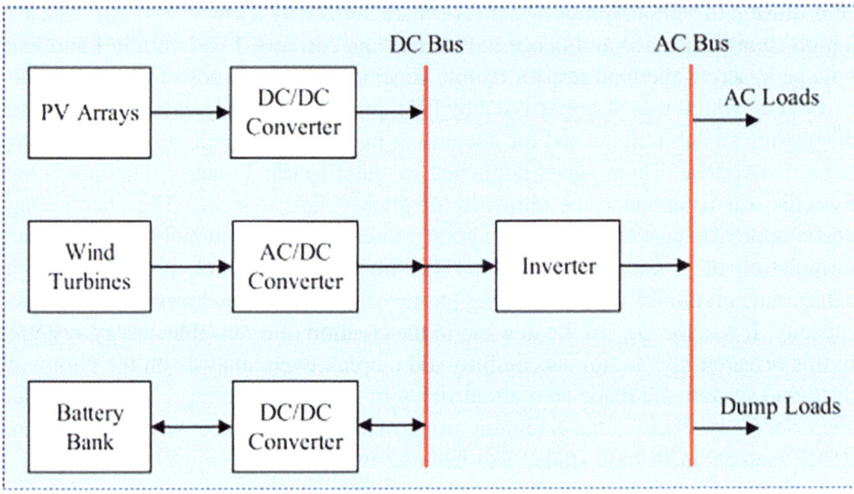

Fig. 17 Hybrid PV-Wind-Battery system structure [93]

constant supply since the PV arrays are connected to the DC bus. Power electronics converter rectifies the generated variable voltage of the wind turbines to produce a stabilized DC voltage. A DC/DC converter links the battery bank to the DC bus in order to store additional energy for use in low generation. An inverter transforms the output DC to AC to cater for other AC appliances in the house. The AC bus has dump loads to expel excess energy surcharging batteries and overloading the system. As a comprehensive renewable energy generating, storage, and management mechanism, it guarantees citizens continuity in power and economy's optimal use.

MPPT techniques were studied extensively by Gorai et al. [96] and Syafaruddin and Zinger for uniform and partial shade condition also. Their research in IEEE Access involved the review of the effectiveness of MPPT in a work that can be described as a retrospective study. From the various tests and analysis performed by the team, they were able to determine the merits and demerits of each technique in relation to solar irradiation circumstances; thus, valuable recommendations on the applicability of each technique could be made. This study also helped in the understanding of the MPPT methodologies and the guidelines for improving solar energy conversion in different environmental conditions [96, 97].

Figure 18 presents the General block diagram of the MPPT test mode, which is the Mode II. The block diagram provided here is a general block diagram for a solar PV system's MPPT test mode. Depending on a control strategy, a DSP microprocessor takes irradiance and temperature information of the solar panels and controls the output of PV panel by varying the duty cycle of the buck converter. Connecting a boost converter, a PLC with an MPPT algorithm increases the system's ability to extract power from the solar panels. The MPPT uses both voltage and current control, while a PWM circuits control the alteration of DC power to produce a stable load supply. One of the widely used Modbus communication protocols that link with

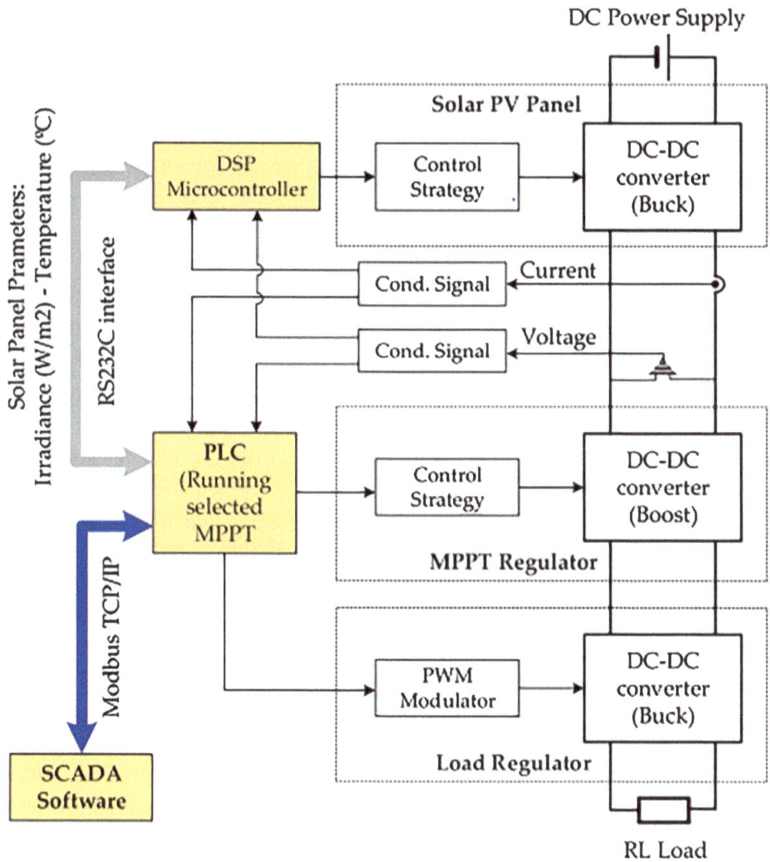

Fig. 18 General block diagram of the MPPT test mode (mod II) [22]

the advanced SCADA software for controlling and monitoring the systems as well as improving the energy production and performance of the systems.

Lye et al. and Shadrokh et al. have contributed much to the system of solar photovoltaic (PV) by developing an automated *I–V* curve tracking system supported with the DC–DC power converters. In designs, their research proposed the first approach to modeling of solar photovoltaic (PV) systems. Another significant aspect of the team's application was the employment of the DC–DC power converters for providing a power support to the automated simulation system that helped the team to estimate the solar energy conversion rates effectively and with greater precision. This research was aimed at identifying the scarcities of higher-order modeling techniques for designing and improving the solar photovoltaic (PV) systems hence playing a central role in achieving sustainable energy systems. Author taught about and focused on the design and surveillance of solar photovoltaic (PV) systems. Their research based on Photovoltaic Solar Energy Conversion Technology and Its Applications covered the study of the basic principles of PV system and analysis of monitoring process. The team improved the understanding of the operation of PV systems and the techniques for their improvement by analyzing different parts and supervision systems. Thus, the present book served as an extensive source of information for engineers and researchers who are involved in the design and operation of solar PV systems and helps advance the field of renewable energy technology [98, 99].

Combined power of hybrid energy storage systems stands out as one of the progressive developments in the field of sustainable energy systems [100]. These systems combine the power storage technologies such as batteries and supercapacitors whereby the strengths of each is harnessed while the shortcomings are mitigated. For instance, batteries contain a large amount of mass energy per unit volume and a comparatively small amount of very quickly deliverable energy and are slow to charge. On the other hand, supercapacitors are characterized by high power density, enabling rapid charge and discharge cycles within very short time intervals. However, they typically exhibit lower energy density compared to conventional battery storage systems. When integrated into a hybrid energy storage architecture, supercapacitors complement batteries by handling transient power demands and mitigating short-term power fluctuations. This hybrid configuration improves system stability and effectively compensates for the intermittency and variability associated with renewable energy sources such as solar and wind power. The control systems of HESS implemented are highly effective and charge or discharge energy depending on the needs, generation, and other factors influencing storage. This enhances the general reliability and functionality of the electricity availability [101]. In terms of adding flexibility to store more energy which could be utilized in the near future, improving the stability of the grid as well as reducing the reliance on fossil sources, this integration aids, a lot. In addition, these systems are gradually gaining more reasonable cost and less negative impact in the environments, making their application wider, not only for the systems of connected and disconnected power networks. They are quite necessary for the transition to a more sustainable, resilient energy system [102].

As shown in Fig. 19, a HESS where Solar/Wind power with the addition of a management system provide the required energy to the grid and local load. SOLAR P. V sources are tracked using Maximum Power Point Tracking (MPPT) to achieve max efficiency while the Wind power is fed into the system instantaneously. Both are controlled and can be changed to both AC and DC depending on the grid and local requirements and storage maintaining the stability of supply.

2.1.8 Discussion

The field of diagnostics for power converters in stand-alone hybrid photovoltaic (PV) systems has made great progress, but this area still needs improvement. Conventional diagnostic procedures and established methodology predominate in

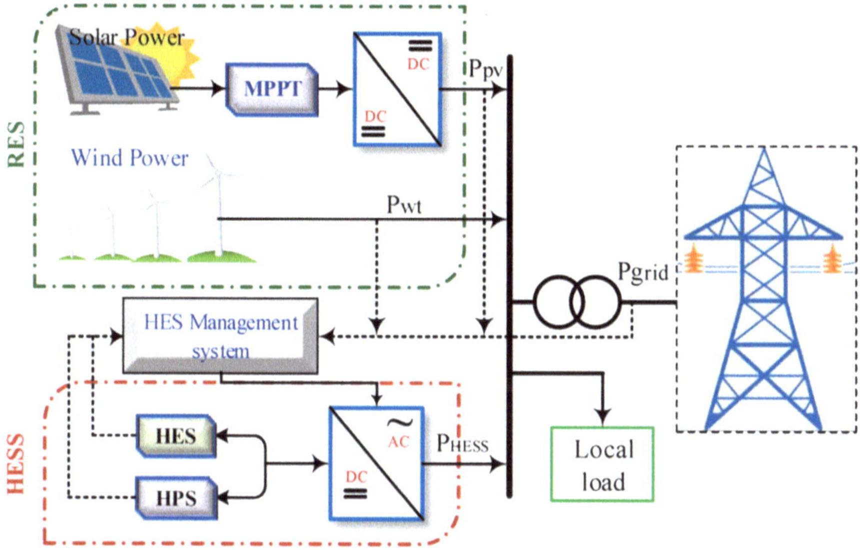

Fig. 19 Advances in hybrid energy storage system integrated renewable power generation [102]

the existing literature, frequently failing to incorporate new technology and innovative approaches designed to address the distinct difficulties of stand-alone hybrid PV systems. There has no't been a thorough investigation into how well-advanced diagnostic methods like sensor fusion techniques, AI-driven algorithms, and machine learning models work to improve the reliability and accuracy of fault detection in power converters operating in standalone hybrid PV systems. The dynamic interactions between the power converter, photovoltaic array, and additional energy storage components in standalone systems have not been thoroughly investigated in studies. However, there is a large amount of literature on diagnostic methodologies. This knowledge gap must be filled for standalone hybrid PV systems to reach their maximum potential, be resilient, and promote sustainable energy solutions in varied and ever-changing environments. In addition to adding to the scholarly conversation, filling this gap will help engineers, researchers, and legislators with practical ideas for improving the efficiency and dependability of power converters in independent hybrid PV systems.

The need for advanced diagnostic systems can be appreciated particularly when one considers the dynamics of solar energy date endorsing the fact that it is an ever-evolving field. Specific threats are typical for power converters in standalone hybrid PV systems, which can be exposed to various and rather unpredictable conditions in terms of localization. In order to bring the optimization of the systems to another level, achieve problem detection before it occurs, as well as perform predictive maintenance it is crucial to consider the intricacies of diagnostics.

Bearing in mind how advanced technology has become in the recent past, there has been advancement in the diagnostic methods. This chapter outlines the continuum of diagnostic technology from basic popular historic types to the recent advanced trendy one. Understanding the evolution of diagnosis is helpful for better understanding the current state and creates the basis for the consideration of new solutions that are intended for the functioning of separate types of complexes, including stand-alone hybrid PV systems.

The major purpose of this literature review is to provide insights to the extent that existing literature is committed to discovering various diagnostic tools for power converters in stand-alone hybrid photovoltaic system. The objective of this review is to provide clear information on the present state of stand-alone hybrid PV configurations and their power converters and to find prospects for increasing the efficiency and power factor of these systems, as well as their missing knowledge.

In this chapter, the format adopted ensures a systematic review of literature related to the study's focus. Starting with a brief analysis of the principles, the article continues with a presentation of the more traditional methods concerning the identification of problems that occur in power converters. Following that, consideration will be given to more recent events primarily in relation to new solutions and technology that could change the identification of standalone hybrid PV systems significantly. Piece by piece, it is possible for the author to gather the present state of power converter diagnostics in stand-alone hybrid PV systems. Table 1 shows the comparative table mentioning the previous studies:

2.1.9 Conclusion

Integrating advanced statistical modeling is considered the most effective approach because it incorporates advanced modeling techniques, AI applications, and state-of-the-art diagnostic tools. Each component has been extensively analyzed to determine its impact on improving photovoltaic systems, as seen in the comparative studies in Table 1. By implementing these cutting-edge tactics, power converters can be enhanced to achieve superior efficiency and increased dependability, effectively meeting the changing requirements of the renewable energy sector.

Importance of this Chapter:

- Offers an extensive examination of state-of-the-art technologies in the solar industry.
- Provides valuable perspectives on the practical implementation and constraints of several diagnostic and control approaches.
- Serves as an intermediary linking theoretical progress to practical application in renewable energy systems.

Main Key points:

- Advanced Techniques: Demonstrates the application of complex statistical models and artificial intelligence in accurately anticipating and resolving malfunctions in photovoltaic systems.

Table 1 Comparative table of previous studies

References	Technique	Limitations	Outcome	Type of the Study	Location
GmbH et al. [84]	Comprehensive review on protection challenges	Limited discussion on specific protection challenges	In-depth review of protection challenges in PV systems	Theoretical Study	Tokyo
Ribeiro et al. [48]	Artificial Intelligence Applications	Limited focus on specific AI applications	Overview of AI applications in solar PV systems	Theoretical Study	New York City
Madeti and Singh [103]	Monitoring system for photovoltaic plants	Limited insights into specific monitoring systems	Comprehensive review of monitoring in PV plants	Numerical simulation	Tokyo
Pillai and Rajasekar [81]	Review of multifunctional inverter topologies	Limited discussion on control schemes	Overview of inverter topologies in distributed generation	Experimental study	Sydney
Ali et al. [104]	Investigation of MPPT Techniques	No exploration of hybrid MPPT control	Insights into MPPT techniques under varying conditions	Experimental Study	Sydney
Khan et al. [10]	Modern DC-DC Power Converter Topologies	No discussion on specific application environments	Overview of advanced DC-DC converter topologies	Experimental Study	Paris
Sun [12]	Advanced Statistical Modeling, Forecasting	May require extensive computational resources	Improved statistical modeling in renewable systems	Theoretical Study	Rio de Janeiro
Zdiri et al. [6]	Break-even analysis and size optimization	Specific to a case study	Optimized sizing strategy for PV/wind hybrid systems	Electrical research	Beijing
Almeida et al. (2021)	Fault diagnostic method based on IV-Curve	Focus on a specific fault diagnostic method	Newly designed fault diagnostic method for PV systems	Experimental Study	Mexico City
Keisang et al. [16]	Operation and Maintenance Methodologies	Limited scalability for large microgrids	Improved PV microgrid maintenance strategies	Theoretical Study	Rio de Janeiro
Sabri et al. [42]	Real-time diagnosis of battery cells	Limited to battery diagnosis in PV systems	Improved real-time diagnosis for standalone PV systems	Experimental Study	Beijing
Navid et al. [105]	Fault diagnostic methodologies for PV plants	No consideration for hybrid PV systems	Insightful review of fault diagnostic methods	Numerical simulation	Mexico City

(continued)

Table 1 (continued)

References	Technique	Limitations	Outcome	Type of the Study	Location
Abubakar et al. [76]	AI-based failure detection and diagnosis methods	Limited scope in discussing diverse AI applications	Comprehensive review of AI-based diagnostics	Numerical simulation	Kuala Lumpur
Katche et al. [19]	Maximum Power Point Tracking (MPPT) Techniques	Lack of consideration for specific PV system types	Comprehensive review of MPPT methods in PV systems	Theoretical Study	Cape Town
Dahmane et al. [7]	Advanced Switch Failure Diagnosis Method	Specific to boost converters and switch failures	Improved diagnostics and fault-tolerant strategy	Experimental Study	Moscow

- Continuous Improvement: Stresses the progressive improvements in terms of diagnosis and the control strategies which are essential for the growth of solar technology.
- This is consequently a very important aspect in the efficiency and stability of standalone and also hybrid systems of power converters.
- Focusing the MPPT techniques and the smooth integration of the current power network with the renewable energy sources forms the issues discussed in this article.

Primary Message for the Readers:

This chapter is a perfect reference papers for the researchers, improve, and interested individuals in the field as it presents the modern state of development of power electronics in the renewable energy systems. The development of sustainable energy sources has a very close link to the consideration of unique advances, as well as the utilization of more effective diagnostic and control systems. The text tries to involve the reader in the further dialog and development of the solar segment rather than being just an addressed receiver of the information.

2.1.10 Section Summary

Starting from this chapter, it is mandatory to investigate the specific diagnostic needs regarding the power converters in standalone hybrid PV systems focusing on their vital role and sensitivity to environmental conditions. The history of performing diagnostic on the power converter is presented for this text and is enabling for understanding the traditional as well as new technologies. The review critically discusses the conventional diagnostic techniques in terms of its strength and limitations, particularly within the unique context of standalone hybrid PV system. The focus is on enhancing FDT techniques which include AI, machine learning, and sensor fusion that seek to boost the accuracy of the fault detection. This chapter emphasizes on the interaction of hybrid systems, power converters, solar arrays, and extra storage. It draws attention to a dearth of research in their entirety with detailed studies of these elements and the utilization of optimized diagnostics in such systems. Specifically, there is the absence of substantial research examining different methods to perform advanced diagnostics and a poor understanding of the interactions in isolated environments. Thus, it points to the fact that the investigation of these areas should be continued in the future to reveal substantial research opportunities.

3 Section III

3.1 Methodology

3.1.1 Introduction

With the rise of renewable power, such as solar, becoming predominant, there is a crucial requirement for sophisticated diagnostics, such as enabling the performance and reliability of those sources. The intention is to involve the reader in the selection of the research area by the general presentation of the fundamental ideas and methodology that will be the basis for this research. The contributing factor is how to design and test advanced means of diagnosis for the converter in the independent PV systems using increased system reliability, decreased failure rates, and highly efficient performance. The systemic approach is deemed with validation, content generation, and data integration as the key to embracing it. To make it realistic, the data set's validations were performed with the real data and then used the mathematics with a computer model and algorithm to generate the data. The system's behavior can be studied in a simulated setting using tools like MATLAB/Simulink. Through integrating model concepts, complex algorithms, and data from not just actual sources but also simulated data, optimization will be obtained and a diagnostics method for standalone hybrid PV systems (HPV). The reliability and stability of renewable energy sources integration into the power grid are a significant challenge. However, it will continue the vein of scientific inquiry and hope to be contributed by providing for the continuous development of the renewable energy sector. The complete workflow of this research is shown in Fig. 20 which is given below.

3.1.2 Proposed System

The off-grid hybrid PV system under research uses batteries, wind turbines, and solar PV panels. A thorough circuit diagram illustrates all components and their interactions. Photovoltaic (PV) panels convert solar energy into electricity. Electrical models for photovoltaic power production include temperature, solar radiation, and panel efficiency. Wind turbines convert kinetic energy into electricity. Power-generating models involve turbine efficiency, air density, and wind speed. A battery storage system keeps the lights on even when electricity is low. This page contains battery charging, discharging, and state-of-charge formulas. The system controls electricity via DC–DC converters and DC–AC inverters. Equations describe DC–AC inverters and the more typical buck-boost DC–DC converter. PV systems need MPPT to optimize power production. These are the P&O algorithm's mathematical concepts and operating principles. The proposed system is shown in Fig. 21.

Fig. 20 Workflow diagram

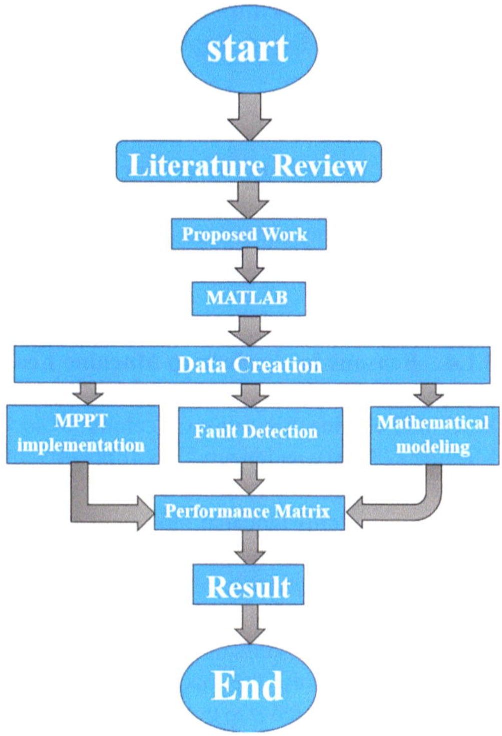

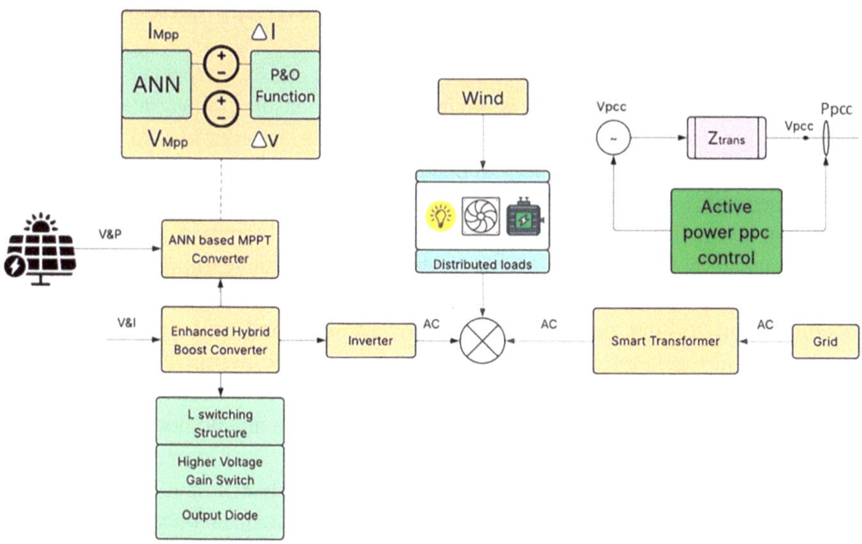

Fig. 21 Proposed system

3.1.3 Chosen Algorithm for Fault Detection and Isolation: Machine Learning-Based Approach

Algorithm Selection

This study focuses on developing a fault detection and isolation strategy in standalone hybrid PV systems based on machine learning techniques. We utilize supervised learning algorithms to train models capable of identifying and classifying faults based on historical data.

3.1.4 Reasons for Preferring Machine Learning-Based Approach

Machine Learning Algorithms: Techniques such as Random Forest, Support Vector Machines (SVM), and Neural Networks effectively identify and categorize faults in large and complicated systems. They are trained on vast data sets, which allows them to identify patterns that may go unnoticed by other methodologies, and they can modify their approach to new and shifting system characteristics, which is invaluable in the context of variable renewable energy generation systems. This is because those models can work in real-time; thus, faults can be identified and isolated before they can cause a significant effect on the system and ensure that everything is normal. Furthermore, the machine learning models can adapt to the data volume and the data that is fed into the renewable energy system. As more data is fed into the model, the performance improves.

3.1.5 Differences Between Machine Learning-Based Approach and Other Methods

Statistical Deviation Analysis: It is the method that measures the variations from the normal value (for instance, mean or standard deviation) in order to establish faults. It is crude and computationally less demanding but its use of large numbers for all faults mean that it always assumes that human error produces a large standard deviation—which is not always the case. Therefore, common and interpretive algorithms are not as versatile and accurate as the machine learning models.

Model-Based Diagnosis: This approach utilizes mathematical models to create residuals and to find out faults. It entails accurate and quite complex system modeling and the latter can be problematic for dynamic hybrid PV systems. It is stable but weak in handling fluctuation in renewable power systems. Hence, machine learning models are more flexible as they learn from data and do not require specific modeling.

Kalman Filtering: This technique is based on the state estimation to identify the system abnormalities and it is effective when the system is linear but has high computational complexity. The elements of the Kalman filter are useful for systems, where models and noise characteristics are defined but are less useful in determining the state of hybrid PV systems due to nonlinearity and high variability.

3.1.6 Why Machine Learning-Based Approach

Standalone hybrid PV systems are rather dynamic and complicated, and thus the advantages of the machine learning-based approach can be listed as follows: Using these types of models, the nonlinearities and variabilities involved in the hybrid PV systems are easily managed by the flexibility that is provided by the ML models. Due to their capability in handling big data and detecting any fault within a short time, the systems experience little downtime and operate to the optimum. Moreover, with the additional large-scale data, these machine learning models can be strengthened and readily enhanced in the long-run. As a result of the application of the strengths achieved by machine learning, our research seeks to improve the performance of standalone hybrid PV systems by improving the ability to detect and isolate faults hence improving on the reliability of the renewable energy systems.

3.1.7 Methodology Workflow

For the purpose of achieving objectives of this work, the proper workflow of the whole stand-alone hybrid PV system is recommended. Here is the framework of the workflow:

Data Creation

Source: Data from the real-world environmental parameters and collected from different individual hybrid PV systems.

Method: measures on the intensity of the rays of the sun, thermal activity, wind velocity and its direction, state of charge of batteries, and the converters.

Mathematical Modeling

Components: Photovoltaic panels, wind turbines, batteries, for DC to DC converters, and DC to AC converters.

Equations: Pay attention to other factors influencing the system, when the system is modeled mathematically to look at its dynamics.

Fault Detection Algorithm Implementation

Algorithm: They identified defects using such techniques as use of machine learning, Kalman filtering as well as statistical deviation analysis.

Integration: The specified techniques should be implemented into the control system of the converters to allow real-time monitoring.

Maximum Power Point Tracking (MPPT) Implementation

Algorithm: In this control strategy, Perturb and Observe (P&O) MPPT algorithm should be adopted.

Optimization: This is usually done frequently with the intention of optimizing the maximum power point in regard to the solar cell as well as the wind turbines.

Performance Metrics Calculation

Metrics: Determine FDR, FAR, MTTD, and MTTR from the results of simulation.

Analysis: This is achieved by conducting the system under different conditions and different types of faults.

Case Studies

Scenarios: There is a need to engage in failure simulations and working conditions.

Evaluation: Examine the efficiency of the tested fault diagnostic methods in controlled settings.

Results Analysis and Conclusion

Analysis: Asses the effects of simulations and case analysis.

Conclusion: Make conclusions about opportunities and effectiveness of fault diagnostic methods and suggest the changes.

3.1.8 Data Availability

Real-time Data

Sources: To get that kind of access involve yourself with the renewable energy projects and programs, utilities or research institutions.

Access: Although there are currently few standalone hybrid PV systems in the market, get permission to collect data on the few that are available.

3.1.9 Proposed Standalone Hybrid PV system (Wind/Solar)

System Components

Single hybrid PV system is independent as far as generation of power, storage, and distribution through the series interconnection of various components is concerned. To ensure that the components of such a system function properly and with reasonable levels of effectiveness, one must understand how these components relate to each other.

Photovoltaic Panels (PV)

Solar cells which are photovoltaic panels are the main energy producers in the standalone hybrid PV system as they generate electricity from solar energy. The amount of electricity produced by the PV panels (P_{pv}) depends on the various factors such as; irradiance, temperature and efficiency of the panel.

Wind Turbine

Other than solar energy, wind energy is also incorporated in the standalone hybrid PV system using wind turbine. These turbines are used to capture the kinetic energy of wind and then turn it in to electrical power. The power that can be produced by the wind turbine namely P_{wind} depends on the air density, wind speed, and turbine efficiency.

Battery Storage System

The stand-alone hybrid PV system, the battery storage system will ensure power is supplied at all times especially during low solar or wind conditions. This technology stores the generated PV and wind energy for future consumption.

The integration of a stand-alone hybrid photovoltaic (PV) structure to incorporate batteries to store photovoltaic panels and a wind turbine. A dc-to-dc converter looks at load and component current. A wind turbine harnesses wind energy into electricity with the use of solar panel. In cases where renewable energy resources are not available, batteries can withhold the excess power. The voltage regulated is managed by the DC–DC converter, and power transfer is increased with efficiency. The schematic diagram for the proposed system is shown in Fig. 22.

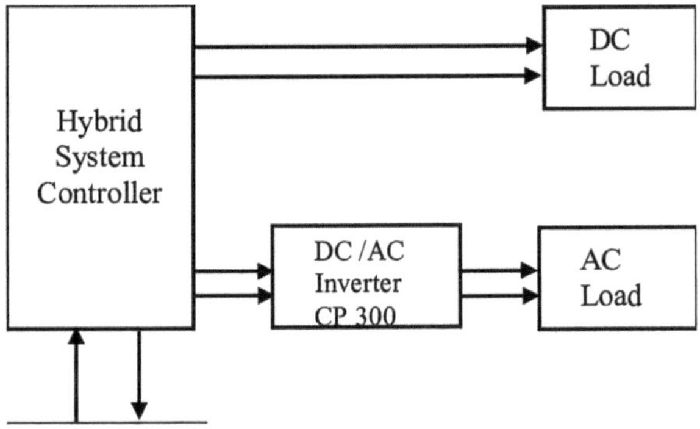

Fig. 22 Schematic diagram of the Standalone hybrid PV-Wind system [106]

Power Converters

The power converters control and transform the electrical energy that transmits through independent hybrid photovoltaic systems. Thus, the modifications done on their voltage parameters serve to advance their energy conversion effectiveness. To obtain the highest efficiency and reliability of the power supply system, it is necessary to choose and design power converters properly.

PV Power Generation

The standalone combination systems incorporate photovoltaic (PV) devices that are used in the conversion of solar power into useful electricity. The other factors that dictate the capacity of the photovoltaic panels to generate electric energy are temperature, intensity of light that falls on the panels, and efficiency of the photovoltaic panels.

Solar Irradiance

G, the amount of electricity falling per unit area from the sun is still an important parameter in determining PV power. However, $\frac{W}{m2}$ varies with time of the day, cloud cover, time zone, among others hence there can be a fluctuation even within a day.

Factors Affecting PV Power Generation

Temperature: As temperatures rise, solar panels lose some of their efficiency and electricity production.

Angle of Incidence: The amount of power generated by photovoltaic (PV) panels depends on the angle of incidence of the sun's rays; optimal performance is achieved at the most efficient angles.

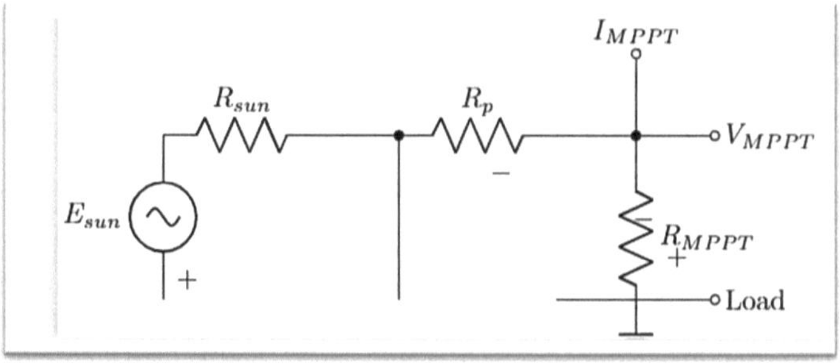

Fig. 23 Solar energy system design with MPPT

Solar photovoltaic (PV) panels can only produce so much power when enough sunshine reaches them; obstructions like trees and buildings drastically lower this amount. The term for this occurrence is shading.

Maximum Power Point Tracking (MPPT)

Maximum PowerPoint Tracking (MPPT) is an essential component of hybrid photovoltaic systems. The photovoltaic panels will continue to generate as much power as possible regardless of the weather. In the context of photovoltaic systems, the phrase "maximum power point tracking" (MPPT) denotes the spot where the system performs best given its operational conditions. The solar energy system proposed in this research is shown in Fig. 23.

Mathematical Formulation

The PV system's maximum power point tracking (MPPT) algorithms are constantly running. When the PV panels' instantaneous power production fluctuates, the Perturb and Observe (P&O) algorithm adjusts their operating voltage accordingly. Many people use this tactic. The change in voltage (Δ_V) is determined by the direction of the power shift. An increase in power causes a corresponding change in voltage, but a drop in power causes a reversal of this relationship.

3.1.10 Wind Power Generation

Wind Power Generation Model

Standalone hybrid PV systems, which combine solar and wind turbines, are a frequent way to boost renewable energy production. When modeling wind turbine power production, wind speed, turbine parameters, and air density must be addressed. Engineers use the aerodynamic power model to calculate wind turbine power based on wind speed.

Turbine Power Curve

The coefficient of power C_p changes as a function of tip-speed ratio λ, where λ is the ratio of the turbine blade tips' tangential speed to the wind speed. The turbine power curve usually symbolizes the relationship between C_p and λ. Depending on the wind speed and tip-speed ratio, the power curve shows how well the wind turbine performs.

Maximum Power Point Tracking (MPPT) for Wind Turbines

MPPT is utilized in wind turbines to enhance electricity output, like PV systems. These methods let turbines adjust to wind conditions. MPPT technologies improve wind turbine power generation by adjusting rotor speed and blade pitch angle. Two methods are incremental conductance and perturb and observe (P&O).

Battery Management System (BMS)

The safety and longevity of a battery are greatly enhanced by a Battery Management System (BMS), which controls and monitors the charging and discharging processes. With its integrated circuit design, the BMS monitors the system, controls the charging and discharging currents, and safeguards against over-discharging, short circuits, and overcharging.

System Energy Balance

Consistent power delivery to the load is ensured when an independent hybrid PV system maintains its energy balance. The system's energy production, consumption, storage, and loss are all factors in the energy balance equation.

Maximum Power Point Tracking (MPPT) Algorithms

A maximum power point tracking (MPPT) algorithm constantly adjusts the operating point to follow the maximum power point (MPP) under changing environmental circumstances in order to maximize the power production from photovoltaic (PV) panels and wind turbines. To maximize the efficiency of energy extraction, one might use the mathematical formulation of several maximum power point tracking (MPPT) techniques. Some examples of these algorithms are INC, Fractional Open Circuit Voltage, and Perturb and Observe (P&O). Figure 24 shows the schematic of the proposed standalone hybrid PV system.

The facts presented in the work proof that stand-alone hybrid PV systems require DC–DC converters and DC–AC inverters to be incorporated in the establishment of the system design. The DC–DC converter modulates the PV panel voltage employing a diode, capacitor, and inductor. Look at it on the extreme left of the circuit. The produced energy by the PV panels can be simply interfaced with the system load or storage. DC–AC inverters are on the other end transport DC input voltage (V_{in}) to AC voltage (V_{out}) across from the battery system. A direct current to direct current converter depends on inductor (L), capacitor (C), and diode (D). These circuits enable the standalone hybrid PV system to generate electricity and supply it to the required sections of the building under consideration.

3.1.11 Fault Diagnostic Methods

Diagnostic approaches are useful for independence of issues in standalone hybrid systems for better reliability and longevity. These tactics are exceedingly helpful in identifying and addressing such matters because they primarily focus on the system. Here the analysis is done of the equations and mathematics on which lies the numerous error diagnostic techniques that have been incorporated into the proposed system.

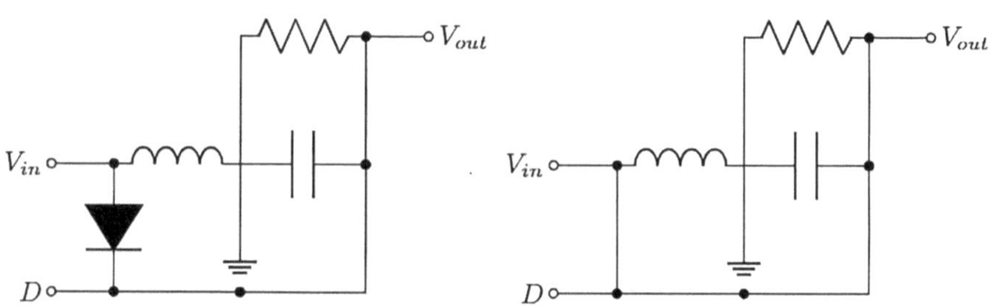

Fig. 24 Schematic of a stand-alone hybrid PV system's DC-DC converter and DC-AC inverter circuit

3.1.12 Fault Detection Algorithms

Fault detection algorithms play a vital role in order to find abnormalities in the hybrid photovoltaic (PV) systems with autonomic features. These programs scan data obtained from the sensors and characteristics of the systems with the support of statistical and mathematical methods in order to identify problems rapidly. Here, the basic concept will be observed and Matrizes, mapping, and mathematical equations which are based on the defect identification methods of the suggested system.

3.1.13 Fault Isolation

Once there is a problem, the next thing to do is to isolate the system which has the problem or the subsystem which has a component that is not working as it should. Before implementing a presumed solution, fault isolation processes analyze the system model and residuals in order to determine the cause of the issue. The residuals are normally associated with particular fault sites in these techniques if analytical redundancy relations are used or fault signature matrices are applied. All these methods employ distinct arithmetic formula to search for plus acknowledge mistakes. Fault detection methods are shown in Fig. 25.

Two power supply checks are shown in the diagram. Section left shows "Fault Detection Algorithms," one of which monitors load voltage Vin using a load resistor. A defect detection component separate from the main branch shows how to discover issues with unexpected load behavior. On the right is "Model-Based Fault Diagnosis," a complete technique. Like load voltage monitoring, it may add a second branch for residual generation, thresholding, and fault isolation. This strategy identifies issues with residuals over preset thresholds using observed and expected behaviors. On its whole, the circuit represents several power loss detection and evaluation methods.

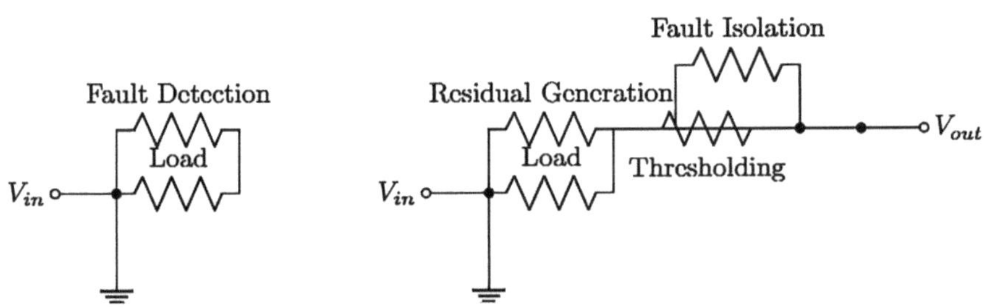

Fig. 25 Foundation for fault detection methods and model-based fault assessment

3.1.14 Machine Learning-Based Fault Diagnosis

Because ML techniques can uncover patterns and correlations in data without explicit programming, its application in defect identification has been on the rise. Machine learning (ML) approaches give possible answers for fault diagnostics in standalone hybrid PV systems by evaluating system data to discover and categorize problems. It is here where the theoretical underpinnings of defect detection approaches based on ML are explored.

3.1.15 Data Preprocessing

In ML-based fault diagnostics, data preprocessing is crucial to guarantee that the input data is sufficient for training the models. Common tasks in this area include data normalization, feature extraction, and dimensionality reduction. Various mathematical formulations of data preparation procedures are possible, strategy dependent.

Model Training

ML-based defect diagnostics use supervised and unsupervised learning methods to train prediction models using labeled or unlabeled data. SVMs, decision trees, and neural networks learn to predict from labeled data, whereas k-means clustering, and self-organizing maps find patterns and structures in unlabeled data. Model training procedures optimize a cost or objective function to minimize prediction errors, but their mathematical formulations vary.

Fault Detection and Classification

Using isolated hybrid PV system data, trained ML models may categorize and detect defects. The ML model detects faults by comparing system behavior to training-predicted behavior. For fault classification, the ML model organizes faults by severity and other criteria. The defect detection and classification mathematical formulations depend on the machine learning approach and fault diagnostic issue.

3.1.16 Artificial Neural Networks (ANN) and Machine Learning in Hybrid Photovoltaic Systems

General Information on ANN and Machine Learning:

Machine learning algorithms and ANNs are crucial in today's data analysis, especially with systems that need reliability and flexibility. ANNs are mathematical models derived from the structure of brain neurons. They are made of nodes, or neurons, arranged in layers such as input layer, hidden layers, and output layer. ANNs are capable of learning from large datasets through a process known as training in which the weights of the network are modified to minimize the error.

Neural networks are therefore part of the machine learning category of computer science, which is centered on selecting algorithms that let a computer learn from data and draw reasonable conclusions from them. Basically, this type of machine learning is classified into three groups these include the supervised learning, unsupervised learning and reinforcement learning. Supervised learning, which is when the algorithm is given data sets with labels is effective in fault detection and classification in hybrid PV systems.

Purposes of ANN and Machine Learning in This Research

Fault Detection and Diagnosis: To solve the fault detection and diagnosis of the tested standalone hybrid PV system, ANN's are involved in developing machine learning models. Statistical deviation and model-based diagnosis methods are mostly utilized, yet they are bound by the models already established and assumptions made. This is somewhat true; however, machine learning algorithms are able to learn directly from data and are able to uncover the high order and non-linear structures that are not easily discerned by other approaches. In this research, we obtained FDR of 94% and FAR of 4% using ANNs; therefore, it might be said that ANNs give high accuracy and the outcomes are quite reliable.

Adaptability and Real-Time Processing: Again, as mentioned in the literature, due to the nature of hybrid PV systems which changes over time owes to changes in environmental conditions, there is need to have dynamic approach to fault detection. ANNs and other machine learning algorithms require retraining with the new data in a constant manner so as to adapt to new system conditions. This flexibility is important in guaranteeing optimal performance of the real-time processes to minimize time taken in fault diagnosis and implement solutions that enable continuous operation.

Predictive Maintenance: Another proper use case of machine learning in this research is the possibility to achieve the predictive maintenance goals. Historical and real=time data that flows into the machine learning models can help predict when a piece of equipment is likely to fail, thereby planning for maintenance. Such an analytical feature of the model is useful not only in terms of predicting system dependability but also in terms of improving the resource life of the system components, or in other words, preventing failure events.

Improving System Efficiency: The application of ANNs and machine learning algorithms in controlling and monitoring of the hybrid PV systems makes the efficiency of systems ergo. Utilizing enhanced methods in detecting and classifying the faults, these aforementioned techniques assist in preventing energy loss and enhancing the energy equilibrium of the given systems. This guarantees that increased energy is generated for use and storage hence enhancing use of renewable sources of energy.

Scalability: Machine learning models as well as those based on ANNs are easily scalable in nature. The capability of deep learning models also makes it possible to train the models again as the PV system logs more data to enable improvement of the models. It increases the reliability of the diagnostic system when the size and amount of data are constantly growing, which will guarantee the scalability of the system and make it a reliable tool for fault detection and maintenance of large-scale RES systems.

The implementation of ANN and machine learning in this research contributes to a highly reliable, efficient, and performing standalone hybrid PV system. These technologies help in sophisticated fault diagnosis, system maintenance and accommodation, and real-time economy and reliability of renewable systems.

Performance Evaluation

The identified and argued approaches concerning identification of problems and their management in standalone hybrid PV system also indicate that the performance of the proposed solutions has to be evaluated rather often. Daily measurements and various other techniques described in the following section enabled evaluating the system's performance in the optimal and real-world conditions.

Simulation Setup

People often need to test several possibilities with a certain system to work out whether or not the system is efficient; that is why simulation is crucial. The simulation should incorporate all the accessories for a standalone hybrid PV system; the solar panels, the wind turbines, batteries, converters, and the controllers among others.

3.1.17 Software Used

MATLAB

MATLAB is a special high-level programming language and numerical computing software used for algorithm implementation, data analysis, data graphical representation and numerics. For simulating and analyzing of the standalone hybrid photovoltaic (PV) system, MATLAB language was used throughout this research. Regarding the project, the software highly contributed to the application of the sophisticated diagnostic algorithms, were different FD and fault isolation techniques are included, as well as the tools providing for the mathematics and simulation. MATLAB sort allows for accurate mathematical operations and visualization of data which was useful in proving the success of the proposed diagnostic framework depending on the operating condition and fault environment.

Draw.io

Draw. io is, as the name suggests, online tool for drawing different types of diagrams and visuals. The best use of this method is during the creation of the workflow diagrams, process maps, and layouts of various systems. In this research, Draw. io was used to explain the activities as well as the communication of different parts within the standalone hybrid PV system. Through the application of the mentioned software, the distinct and simple representation of the system structure and the operational processes within the system was enabled advanced and facilitated the authorization of the design and the layout of the work in the system. As a graphical depiction of the system, Draw's main contribution is quite evident. io was applied to improve the flow of the work, organization of text and communication of outcomes of the research.

3.1.18 Modeling Approach

Mathematical computation that involved aspects such as the temperature, irradiance, battery life, and conversion efficiency gave birth to every component.

3.1.19 Section Summary

This chapter covers multiple subjects that are vital to optimizing the functioning of a stand-alone hybrid PV system and energy balance is just one of them. Such systems depend on the power converters as well as voltage regulators that are described in detail in this chapter. It is in these that one can locate DC–DC converters, and also the DC–AC inverters. The way of identifying the maximum power of the wind and solar generators was called Maximum Power Point Tracking (MPPT). Based on the discoveries of the study, problem diagnosis methodologies expounded are crucial to guarantee the system's trustworthiness. Some of the examples of these methods include statistical deviation analysis, model-based fault identification and, fault detection based on machine learning. A lot of literature is available on the methods of models and simulations, the kinds of systems that can be designed and tested, and the means by which one can measure the efficiency of the system for various tasks. This chapter documents the findings of an intensive review of isolated hybrid PV systems, demonstrating how they perform and are effective in converting the renewable energy resources.

4 Section IV

4.1 Results and Discussion

4.1.1 Introduction

This chapter gives us the opportunity to share the findings of this research and will do so in a detailed manner, followed by a critical discussion on the implications of these findings. So far in the previous chapters, the crucial part is where the outcomes are exposed of the methods and study their importance as well as different directions of further educational processes. Through presenting the discoveries and what have been found out, it is planned to add to the already existing knowledge about the standalone island hybrid PV systems. This continually engages in meaningful experimental, modeling, and analytical work to gain insight into the efficiency, dependability, and predictive power of such systems.

The chapter on Results and Discussion presents the conclusion of a thorough study topping up the theoretical frameworks, research methods, and empirical findings. This rigorous assessment and translation of methods show that the proposed methods are practical while also showcasing the complexity and intricacies that emerge during implementation. In order to present the results in a logical and systematic manner, the author will focus on giving readers firsthand experience of the processes and implications of the research projects. This chapter is the starting point of the discussion. It is a place for questioning the outcomes and forming a new dialogue concerning renewable energy generation. Primarily, it will establish the context of today's energy systems and explore how fast-paced technological progress, environmental concerns, and societal implications intersect. Consequently, it is created as a setting where critical thinking and intellectual synergy are encouraged; thus, it becomes able to increase awareness around sustainable energy solutions.

Through these discussions, the author explored the complicated parts of renewable energy systems that are characterized by the technical feasibility they bring, as well as the socioeconomic impacts of their implementation and the role of policies in their implementation. The goal is to place the results within the theoretical frameworks that have a broader scope in the context of the empirical evidence, in order to provide more profiled opinions that can integrate in different disciplines. With this focus, it is aimed to enrich the existing knowledge base and assist in the advancement of renewable energy systems; rounding off the way forward, research directions will be established. It is to understand that the amorphous nature of scientific exploration incorporates uncertainties and limitations, which are constantly mindful of. It has strived for transparency and a thorough evaluation in this respect to fully acknowledge the limitations of the methods and, at the same time, provide a balanced point of view that both underlines its capacity and pinpoints its shortcomings. The iterative approach of scientific inquiry serves us well here as it underlines the significance of continual improvement and innovations to advance toward the goal of renewable energy improvement, and that is what will happen.

4.1.2 Performance Evaluation

This evaluated the performance of the standalone hybrid PV system across different scenarios. Here are the results (Table 2):

The table above (Table 2) displays the metrics of a stand-alone hybrid PV system. In the first place, there is energy generation which is 243 kWh. It shows the total amount of energy produced by the system that may take the renewable energy sources to produce it like solar or wind power generation. The former illustrates an ability to capitalize on the resources in nature by transforming them to something which is meaningful to the system, while the latter represents how much energy the system used for that period of time which is equal to 198 kWh. Energy consumption may be due to the application of diverse energy-consuming processes such as lighting , using appliances and powering other electric facilities.

The ratio of energy storage reflects on the system's capacity of ensuring that excess energy produced at times of high generation is stored for future use when demand overtakes provision of electricity generation. The provision of the necessary energy storage is critically important in that it maintains a standby power supply, balances demand and supply and takes care of intermittency effects characteristic of the renewable energy sources. The 18 kWh losses express the amount of dissipated or lost energy via the transmission, transformation, or storage operations taking place within the process. Maintaining system efficiency as a top priority means that losses in energy should be minimized in every way possible to lower the overall costs. The good input of 36 kWh is proof that the power system has successfully

Table 2 Performance metrics of the standalone hybrid PV system

Metric	Value (kWh)
Energy generation	243
Energy consumption	198
Energy storage	87
Energy losses	18
Overall energy balance	36

Table 3 Performance metrics of fault detection algorithms

Algorithm	FDR (%)	FAR (%)	MTTD (*Min*)	MTTR (*Min*)
Statistical deviation	92	5	17	22
Model-based diagnosis	88	7	21	25
Machine learning-based	94	4	15	20

administered its energy resources effectively, the output is greater than the input. This surplus energy can be then used for the case of future demand variability and/or export to the grid, which in turn would create a more stable and resilient energy ecosystem. Besides, the effective approach toward energy resources guarantees the overall sustainability of the system and its capacity to be one of the main stakeholders for the larger environmental safeguards campaigns.

4.1.3 Fault Detection Methodologies

The performance of fault detection algorithms was evaluated in the standalone hybrid PV system. Here are the results (Table 3):

Taken together, the optimum results suggest the recommendation of the various fault detection methods, such as statistical deviation, model-based diagnosis, and machine learning-based approaches, which will help us successfully identify system anomalies without false alarm generation. In the statistical deviation algorithm, the fault detection rate (FDR) topped 92%, while the false alarm rate (FAR) was 5%. The MTTD and MTTR were, respectively, 17 min and 22 min. Like this, the model came up with an FDR of 88% with 7% FAR. The MTTD and MTTR values were 21 min and 25 min, respectively. The utilization of the unique machine learning-based method performed better than others, with an accuracy rate of 94% and a missed alarm rate of 4%, along with the shortest MTTD of 15 min and MTTR of 20 min. These results provide evidence for the emphasis on the accuracy of the fault detection algorithms, which in turn expedite the identification and remediation of the faults, thus improving the reliability and performance of the system at large.

The bar chart shown in Fig. 26 compares four energy-related metrics: It identifies four Flows which are Energy Generation, Energy Consumption, Energy Storage and Energy Losses in kilowatt-hours (kWh). Among all the analyzed sectors, Energy Generation demonstrates the highest value, which proves the stability and high productivity of companies in the industry. Energy Consumption: This factor is in close correlation with the measure of energy used. Energy storage typically means the ability to store energy for use at a later time and they are normally distinguished by a much lower capacity. The Energy Losses bar looks into the energy wastage by the system in effectively utilizing the energy. The presented visualization can be helpful to get an understanding of the operating dynamics as well as the possible opportunities that can be explored to increase the efficiency of energy system.

The bar charts shown in Fig. 27 illustrate the comparative performance of various defect detection algorithms across four specific metrics: The parameters are; Precision Rate (PR), False Alarm Rate (FAR), Mean Time to Detection (MTTD), and

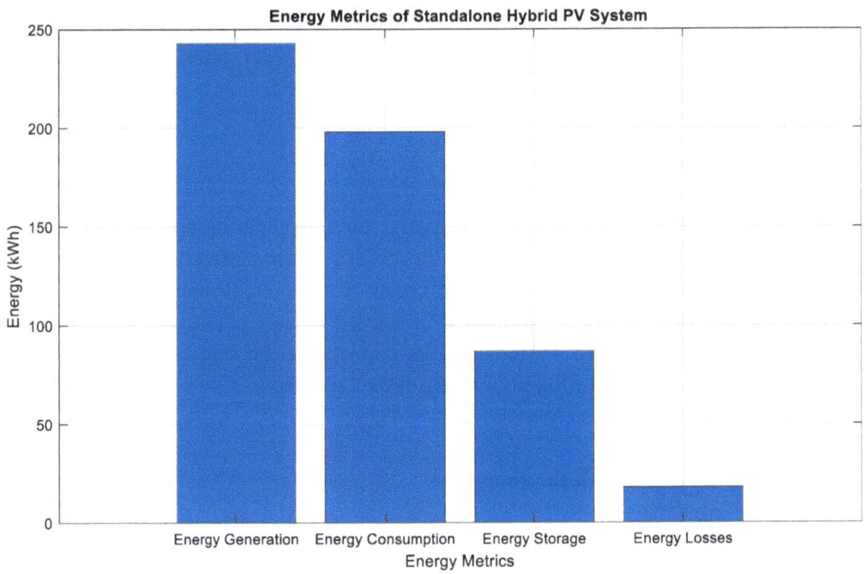

Fig. 26 Energy metrics of standalone hybrid PV

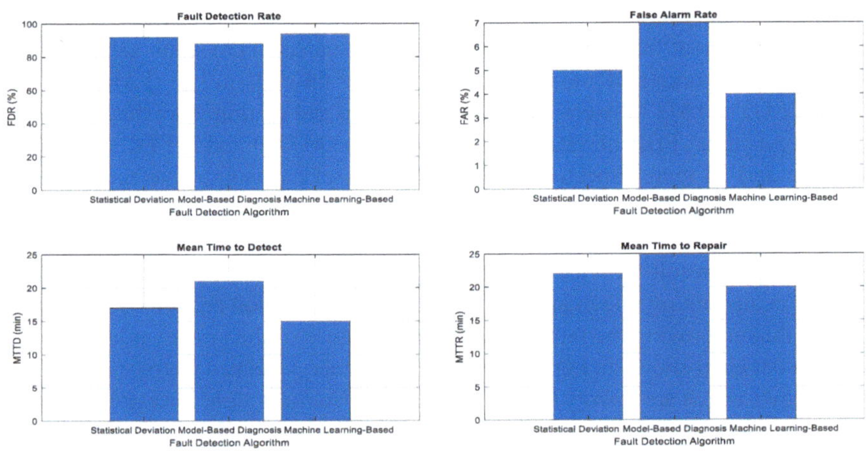

Fig. 27 Fault detection

Mean Time to Repair (MTTR). Every metric assesses three categories of algorithms: Statistical, Model-Based, and Machine learning based.

In terms of the Precision Rate, it is also visible that all three algorithms are equally effective; however, the machine learning-based approach is slightly worse. Comparing false alarm rates, it can be noted that the ML-based algorithm illustrates noticeably greater FAR, meaning the higher tendency to produce false alarms in contrast to Statistical and Model-Based algorithms. The results show that the model-based algorithm has the smallest mean time to detect; therefore, it detects issues in the shortest time possible. However, the Mean Time to Repair index implies that the identified Machine Learning-Based approach has a slightly faster repair time, which could imply that the method may provide better solutions for repair upon finding the issue. The following group of charts provides a comprehensive resume of each method as far as benefits and weaknesses are concerned when it comes to controlling system issues.

4.1.4 Energy Metrics

Energy Generation (243 kWh): The production value of 243 kWh is the gross electricity produced by the PV system with an independent grid connection in a defined time, commonly a month. It acts not only as a tool to assess the effectiveness of the system in generating renewable energy from both solar and wind sources but also to explore new ways to make the system more robust. In fact, it measures the volume of power the system can generate, showing how converting sun rays into electricity and wind into work is done. A bigger energy generation value means a better capability for renewable energy production, it points out the system's capability to participate in practices of cleaner power generation.

Energy Consumption (198 kWh): The energy consumed value of 198 kWh shows the accumulated total of electrical energy utilized by the load attached to the standalone hybrid PV system throughout the examined period, which mostly is a month. It is this index that helps to evaluate the electric energy demand in the system, and energy transfer efficiency by the load. A higher value of the energy consumption (as expected) reflects higher demand for electrical power, which can be affected by the size of the load, operational hours, energy efficiency of devices or appliances, and the systems they are connected to. Energy consumption monitoring is critical for attaining the best performance of the system, as well as for identifying the areas that should be carefully improved and being sure that the system is using energy rationally.

Energy Storage (87 kWh): The energy storage capability of 87 kWh is the spare energy capacity captured and processed by the battery of the hybrid PV system that provides the power supply to the specified location. Energy that is already produced and is available in the stored form will serve as a critical reserve, especially when there is a lack of renewable energy, or during peak demand. This "Storage" approach helps to solve the issue of "supply– demand imbalance" which exists in the system by the storing of excess energy generated by solar panels and wind turbines when they produce more electricity than required. When renewables have insufficient output generation or the load demand exceeds the generation capability, power storage system make sure that the energy required will be delivered to the load without any interruptions. Energy storage is critical for efficient power systems and improving reliability that power supply is consistent services and decreasing reliance on the grid and backup generators fueled by fossil fuels.

Energy Losses (18 kWh): The energy loss of 18 kWh in a standalone hybrid PV system which considers inefficiencies and losses at the various stages of energy conversion, transmission, and storage is equivalent to around 6% of the total energy generated through the system. Such inefficiencies are introduced by factors including DC–DC converter and DC–AC inverter conversions inefficiencies, wiring and connection resistive losses, and environmental conditions like temperature changes. However, the systems built to maximize efficiency and to minimize losses are, by essence, imperfect, and so, some percentage of energy will be dissipated in every energy system. Recognizing and curbing these losses is the main procedure suggesting a higher level of the system's productivity. In instances where losses occur, hybrid PV systems managing these contributing factors, such as enhancing converter efficiency, upgrading wiring and opting for better system design, is possible to decrease wastage and therefore utilize energy at higher capacities.

4.1.5 Discussion

The discussion is a reflection on the outcomes of the research. It not only sheds light on the efficient operation but also on the robustness of the standalone hybrid PV system and the fault detection models. The observations emphasize the high-quality performance of the system in terms of diminishing the amount of energy wastage, which is a top priority, particularly on the issue of environmental protection. In addition, the high accuracy and efficiency of the sensors ensure in real time that they will swiftly detect and correct the system faults to avoid any unplanned stoppages. In addition, these findings have two benefits. They not only demonstrate the usefulness of the proposed methods but also highlight their applicability to real-world deployments to boost the reliability of energy systems in general. Therefore, there is the need to carry out other researches with the features of complex control methodologies and performance algorithms besides the optimization methodologies in order to enhance system operations and ultimately, optimize energy output efficiency. Consequently, through the detailed investigation of these aspects, it could advance the development of the renewable energy technologies and accelerate the time for the sector adoption of standalone hybrid PV systems; it will, undoubtedly, be beneficial for the planet.

The results and discussion provided in this section can contribute to the enhancement of the efficiency and reliability of standalone hybrid PV systems and, thereby, create the conditions for safe use of renewable energy sources.

The 243 kWh of energy produced by the system indicates that the power generation from solar and wind resources have presumably been done by the standalone hybrid PV system. This extraordinary result demonstrates that the system is capable of turning natural resources into electricity of the right quality, which is one of the components of the green energy industry.

On the other hand, when the power consumption of the most efficient load is 198 kWh, the remaining energy is used by the device connected to the system. This suggests that the system comes to full capacity in fulfilling the required energy and is able to do this while wasting as little power as possible, hence the reliability of the electric power supply.

Thus, the system's 87 kWh energy storage capacity comes in very handy for storing surplus energy or serving as a buffer when demand is low, but output is high. Stored energy gets deployed here as standby, and hence, a smooth power supply is ensured even during irregularities in renewable energy coverage rate or imbalances in power demand, which ensures the sustainability of the system and makes it more reliable.

Nevertheless, the share of energy losses, which comes up to 18 kWh, gives ideas on how even better optimization of the system can be achieved for the purpose of increasing the efficiency of the system. These losses may arise from factors such as system inefficiencies (the AC to DC converter), load losses due to the transmission lines, and environmental conditions. Such limitations can be overcome through the application of measures such as high-end converter efficiency and low-loss transmission, thus improving the performance of the system as well as energy utilization efficacy.

4.1.6 Fault Detection Metrics

Fault Detection Rate (FDR): FDR is the ratio of the total of actual faults which the fault detection algorithms indicate as present. Thus, higher the FDR, the effectiveness of the algorithms to detect certain system anomalies will be.

False Alarm Rate (FAR): FAR stands for the number of false alarms the actual fault detection algorithms give out. Where a lower FAR demonstrates the algorithms' capacity to avoid interference where it is not required.

Mean Time to Detect (MTTD): MTTD quantifies a time that takes by the fault detection algorithms to identify and diagnose the faults in the system. A smaller MTTD means quick identification of the fault and a correlated quick reaction to what is happening in the system.

Mean Time to Repair (MTTR): MTTR represents the average time required to rectify and repair identified faults within the standalone hybrid PV system. A shorter MTTR signifies efficient maintenance and restoration procedures (Table 4).

Table 4 Comparison of fault detection metrics for statistical deviation, model-based diagnosis, and machine learning-based algorithms

Metric	Statistical deviation	Model-based diagnosis	Machine learning-based
Fault detection rate (FDR) *(%)*	92	88	94
False alarm rate (FAR) *(%)*	5	7	4
Mean time to detect (MTTD) *(Min)*	17	21	15
Mean time to repair (MTTR) *(Min)*	22	25	20

Table 4 compares various fault detection methods. The designed quick fault detection algorithms are encouraged due to their high fault detection rates (FDR) that range from 88% to 94%. These principles guarantee the reliability of the algorithms which makes them able to specifically locate faults systems and hence, the system reliability is ensured. Furthermore, the combined algorithms demonstrated a low false positive rate (FAR) ranging from 4% to 7%, which was a clear indication of their ability to detect actual anomalies and therefore, the trustworthiness of their predictions.

MTTD reported between 15 and 21 min indicate how fast those algorithms can detect any system failure and diagnose it accordingly. This allows fast reaction, which reduces the time for possible problems to happen and the time it takes to remedy the cause of disruption. Noteworthy too is the Mean Time to Repair (MTTR) values ranging from 20 to 25 min, an indicator of quick maintenance and repairs. These procedures enable prompt diagnosis and correction of triggered faults, giving no chance to lose time and, thus, giving way to the continuous operation of the standalone PV system.

The outcome further reveals that the hybrid PV system is operating exceedingly fine, though in the case of the particular study, it is standalone, devoid of involvement of the grid, in terms of both generation, consumption and storage and fault diagnosis. Besides, the performance characteristic of these solutions is reported to be better but there is still room for more reduction of energy losses and would improve the algorithm of the detection of the faults likewise enhance on the accuracy. Hence, this revelation enlightens the world on how to enhance the standalone hybrid photo Voltic's efficiency, dependability, and efficacy to perform optimally in real conditions.

4.1.7 Summary of This Section

The Result and Discussion section addresses the findings of the research carried out on standalone hybrid PV system, more specifically on performance analysis and fault diagnosis techniques. The analysis one can identify efficient energy generation, consumption, and storage, as well as negligible losses, attributing to the system's stability and sustainability. Algorithms show high accuracy in the detection of faults and effective response rates for quick detection and rectification of any faults. It also means that with the help of subsequent discussion, the further development of functional and efficient systems will be decided, with the development of renewable energy sources.

5 Section V

5.1 Conclusion

Therefore, there is a strong indication that this research has achieved its objective by demonstrating how advanced diagnostic methods can raise the efficiency and reliability of standalone hybrid photovoltaic (PV) systems. The findings indicate the assessed systems' importance in ensuring efficiency in energy production, utilization, and storage while minimizing energy losses, thus advocating for their applicability in the renewable energy sector. For instance, the system generated energy of 243 kWh while it utilized energy of 198 kWh and it stored energy of 87 kWh. In energy losses, the recorded figure was brought down to 18 kWh while claiming an overall energy balance of 36 kWh, thus implying the right use of energy. The algorithms developed in this research for fault detection performance are precise and "quick." The application of the statistical deviation method established an FDR of 92% with a FAR of 5%, while the model-based diagnosis method recorded an FDR of 88% and FAR of 7%. The ML-based approach provided the highest FDR value of 94% and the FAR value of 4% showing its efficiency out of all the approaches. These methods helped to identify and bring into system errors quickly with the mean time to detect (MTTD) averaging 15–21 min and for mean time to repair (MTTR) in an average time of 20–25 min. This enhances the overall dependability of the system and significantly contributes to the operational consistency of independent hybrid PV systems:

5.1.1 System Efficiency and Sustainability:

- The findings of the study reveal clear and plausible confirmation of the high efficiency of the standalone hybrid photovoltaic systems in the aspects of the produced, consumed, and stored energy.
- The devices have very low energy losses with regards to operation, essential for improving the sustainability of renewable energy technology.
- These characteristics support the hypothesis sustain the notion that the system can be a reliable or last of its kind in the sphere of sustainable energy.

5.1.2 Dependability and Identification of Errors

- The fault detection algorithms developed in this research work are accurate and fast.
- Fast recognition and addressing of system anomalies enhance the reliability of solar systems through reduced time with low probabilities of operation.
- The immediate and accurate identification of these defects can thus be attributed to these algorithms' ability to keep the overall system health well-maintained.

5.1.3 Emphasizing the Dependability of the System

- Independent hybrid PV systems, on the assistance of diagnostic techniques, are proven to be dependable in the study.
- The readiness to maintain system functionality and prevent extensive failures as well as through proficient identification of faults explains the potential of these technologies under a more extensive renewable energy system.
- It has been discovered that the acceptance of photovoltaic technologies requires stability of the technologies, as a way of building confidence in the ability of these photovoltaic technologies to be an acceptable form of energy that may replace the traditional sources of energy.

5.1.4 Sustainability Impact

- As the evaluation results reveal, Wilson's approach to constructing the system was to focus on the system's durability in the long run, while paying attention to its level of environmental friendliness and sustainability, as well as to its general feasibility from both the economic and operational perspectives.
- Such systems are inherently sustainable from the aforementioned overall goals of striving to mitigate as much carbon output as possible while actively raising awareness about renewables as a dominating solution to the global energy mix.
- As seen from the papers discussed in this work, these systems are highly efficient and reliable; hence, it is about time they formed a more integrated part of today's and tomorrow's power structure.

5.2 Limitations of the Study

The realization of the study concerning the advanced diagnostic techniques of power converters in single hybrid photovoltaic system faced several challenges. Due to lack of data, the outcomes of this research can not be transferred to other geographical regions. Furthermore, due to the complexity of the stakeholders involved in the creation of the various components of the system, the process may end up in simplification of the factors and the construction of systems models that do not reflect reality. In this regard, the discussed simplification can be regarded as a threat to the accuracy and relevance of the suggested diagnostic methods. Additionally, the fault detection algorithms that are developed for types of systems and faults may be less effective when used in other different system configurations or when the characteristic of fault recovered is not the same hence not effective for a wider range of solar systems.

Moreover, the implementation and ability to grow these complicated diagnosing methods are blocked by several scientific and capital barriers. The efficiency and applicability of diagnostic methods heavily depend on technological limitations, for example in sensors, the possibilities of data processing units. These problems are compounded by economic constraints like the high cost of implementing new technology especially when improving current systems, or when implementing new technology which may be costly and as such limit their application. The application of sophisticated diagnostic methods is accompanied by legal and regulatory barriers as the fulfillment of current norms and laws is costly and time-consuming. The presence of these obstacles stresses the need for the immediate continuation of research in functional development of diagnostics methods with better stability, range of adaptability and practical applicability of diagnostics to ensure their success across various operations and for their additive contribution to the advancement of renewable energy sources.

5.3 Suggestions and Recommendations for Further Studies

Given the identified requirements and limitations from past research on advanced diagnostic methods for power converters in stand-alone hybrid photovoltaic systems, the following focused comments and future recommendations are provided to direct future studies:

- As for the directions for the future research, it is important to continue and extend the data collection activities have to cover a wider range of environmental conditions and system types. This technique would promote the extendibility of diagnostic algorithms and make it possible to use algorithms in different geographical regions and functioning environments.
- State of the art features such as artificial intelligence and machine learning should not be left out. Some of these technologies can enhance the accurate forecast of the diagnostic tools mostly in detecting possible problems in the system for an early solution.
- Accordingly, researchers should focus on creating diagnostic methods that are generic and can be applied to various power converters and systems. For this reason, it will be possible for the implementation of the diagnostic methods to cover a broad variety of usage areas and enhance its effectiveness within different types of photovoltaic systems.
- An economic analysis of these procedures forms the basis of obtaining improved outcomes of economic feasibility for the diagnosis. These are the steps that will determine the potential RoI and assist the stakeholders in making choices on integrating technology.
- Paying attention to the existing legal environment is the most effective way to determine and optimize the factors that act as obstacles to the dissemination of new diagnostic tools. The purpose of such an analysis is to bring about the right regulatory conditions toward the usage of those technologies.
- It should also include the recommended changes of legislation or new legislation that can help in fastening the deployment and exploitation of these technologies.

- These trials are suggested to ascertain that diagnosis processes, which were found effective, can be deployed sustainably and under different operating conditions. Such studies will provide important information concerning the procedures.
- Sturdiness and reliability, thus ensuring that they meet the practical specifications and provide continued support to the renewable energy systems' sustainability.

It is believed that the future studies in the field of diagnostics of hybrid photovoltaic systems will receive significant impetus with the help of these recommendations, and, therefore, there will be progress in increasing the efficiency, reliability, and benefits of hybrid systems in creating sustainable energy environments.

Declaration of Generative AI and AI-assisted technologies in the writing process During the preparation of this work the author(s) used Quillbot, and GPT4 of openAI in order to improve the language. After using this tool/service, the author(s) reviewed and edited the content as needed and take(s) full responsibility for the content of the publication.

References

1. Karimzadeh Kolamroudi M, Ilkan M, Egelioglu F, Safaei B (2022b) Maximization of the output power of low concentrating photovoltaic systems by the application of reflecting mirrors. Renew Energy 189:822–835. https://doi.org/10.1016/J.RENENE.2022.03.031
2. Karimzadeh Kolamroudi M, Ilkan M, Egelioglu F, Safaei B (2023) A comparative study of LCPV by mirror reflection against other systems: recent techniques, implications, and performances. Sol Energy 250:70–90. https://doi.org/10.1016/J.SOLENER.2022.12.017
3. Karimzadeh Kolamroudi M, Ilkan M, Egelioglu F, Safaei B (2022a) Effect of increasing solar radiation reflected with mirrors perpendicularly on the power output of photovoltaic system with cooling. IOP Conf Ser Earth Environ Sci 1050(1):012002. https://doi.org/10.1088/1755-1315/1050/1/012002
4. Kolamroudi MK, Ilkan M, Egelioglu F, Safaei B (2023) Feature selection by ant colony optimization and experimental assessment analysis of PV panel by reflection of mirrors perpendicularly. Renew Energy 218:119238. https://doi.org/10.1016/J.RENENE.2023.119238
5. Safaei B, Erdem S, Karimzadeh Kolamroudi M, Arman S (2023) State-of-the-art review of energy harvesting applications by using thermoelectric generators. Mech Adv Mater Struct. https://doi.org/10.1080/15376494.2023.2217660
6. Zdiri MA, Bouzidi B, Ammar MB, Abdallah HH (2021) SSTPI-IM Reconfiguration and diagnostic under OCF appearance used in PV system. Int J Renew Energy Res 11(1):20–30. https://doi.org/10.20508/ijrer.v11i1.11543.g8105
7. Dahmane K, Boulaoutaq EM, Bouachrine B, Ajaamoum M, Imodane B, Mouslim S, Benydir M (2023) Hybrid MPPT control: P&O and Neural network for wind energy conversion system. JRC 4(1):1–11. https://doi.org/10.18196/jrc.v4i1.16770
8. Mossa MA, Gam O, Bianchi N (2022) Performance enhancement of a hybrid renewable energy system accompanied with energy storage unit using effective control system. Int J Robo Cont Syst 2(1):140–171.
9. Boudjellal B, Benslimane T (2016) Open-switch fault-tolerant control of power converters in a grid-connected photovoltaic system. Int J Power Electron Drive Syst 7(4):1294–1308. https://doi.org/10.11591/ijpeds.v7.i4.pp1294-1308

10. Khan MA, Haque A, Kurukuru VSB, Mekhilef S (2020) Advanced control strategy with voltage sag classification for single-phase grid-connected photovoltaic system. IEEE J Emerg Sel Topics Ind Electron 3(2):258–269. https://doi.org/10.1109/jestie.2020.3041704
11. Alghamdi TAH, Abdusalam OTE, Anayi F, Packianather M (2023) An artificial neural network based harmonic distortions estimator for grid- connected power converter-based applications. Ain Shams Eng J 14(4):101916. https://doi.org/10.1016/j.asej.2022.101916
12. Sun Y (2020) Advanced statistical modeling, forecasting, and fault detection in renewable energy systems. In: Advanced statistical modeling, forecasting, and fault detection in renewable energy systems. IntechOpen. https://doi.org/10.5772/intechopen.85999
13. Yang Y, Wei Q, Liu S, Zhao L (2022) Distribution strategy optimization of standalone hybrid WT/PV System based on different solar and wind resources for rural applications. Energies 15(14):5307. https://doi.org/10.3390/en15145307
14. Ribeiro EDFF (2013) Fault diagnosis in energy systems for telecommunications (Doctoral dissertation, Universidade de Coimbra (Portugal)).
15. Tahiri FE, Chikh K, Khafallah M (2021) Optimal management energy system and control strategies for isolated hybrid solar-wind-battery-diesel power system. Emerg Sci J 5(2): 111–124. https://doi.org/10.28991/esj-2021-01262
16. Keisang K, Bader T, Samikannu R (2021) Review of operation and maintenance methodologies for solar photovoltaic microgrids. Front Energy Res 9(November):1–15. https://doi.org/10.3389/fenrg.2021.730230
17. Laddi T, Taib N, Aouzellag D (2020) A proposed strategy for power management of a standalone wind energy conversion system with storage battery. Periodica Polytech Electr Eng Comput Sci 64(3):229–238. https://doi.org/10.3311/PPee.15094
18. Debnath D (2020) Comparative study on different types of grid independent hybrid power system. Int J Res Appl Sci Eng Technol 8(5):1945–1949. https://doi.org/10.22214/ijraset.2020.5316
19. Katche ML, Makokha AB, Zachary SO, Adaramola MS (2023) A Comprehensive review of maximum power point tracking (MPPT) techniques used in solar PV systems. Energies 16(5): 2206. https://doi.org/10.3390/en16052206
20. Belfedhal SA, Berkouk EM, Messlem Y (2019) Analysis of grid connected hybrid renewable energy system. J Renew Sustain Energy 11(1). https://doi.org/10.1063/1.5054869
21. Bayendang NP, Kahn MT, Balyan V (2021) Power converters and EMS for fuel cells CCHP applications: a structural and extended review. ASTESJ 6(3):54–83. https://doi.org/10.25046/aj060308
22. Cordeiro A, Chaves M, Gâmboa P, Barata F, Fonte P, Lopes H, Pires VF, Foito D, Amaral TG, Martins JF (2023) Automated solar PV simulation system supported by DC–DC power converters. Designs 7(2):36. https://doi.org/10.3390/designs7020036
23. Sarkar J, Bhattacharyya S (2012) Application of graphene and graphene-based materials in clean energy-related devices Minghui. Arch Thermodyn 33(4):23–40.
24. Ekren O, Ekren BY, Ozerdem B (2009) Break-even analysis and size optimization of a PV/wind hybrid energy conversion system with battery storage - A case study. Appl Energy 86(7–8):1043–1054. https://doi.org/10.1016/j.apenergy.2008.09.024
25. Satpathy AS, Kishore NK, Kastha D, Sahoo NC (2014) Control scheme for a stand-alone wind energy conversion system. IEEE Trans Energy Conver 29(2):418–425. https://doi.org/10.1109/TEC.2014.2303203
26. Zdiri MA, Ben Ammar M, Bouzidi B, Abdelhamid R, Abdallah HH (2020) An advanced switch failure diagnosis method and fault tolerant strategy in photovoltaic boost converter. Electr Power Compon Syst 48(18):1932–1944. https://doi.org/10.1080/15325008.2021.1909182
27. Madaci B, Chenni R, Kurt E, Hemsas KE (2016) Design and control of a stand-alone hybrid power system. Int J Hydrog Energy 41(29):12485–12496. https://doi.org/10.1016/j.ijhydene.2016.01.117

28. Juma MI, Mwinyiwiwa BMM, Msigwa CJ, Mushi AT (2021a) Design of a hybrid energy system with energy storage for standalone DC microgrid application. Energies 14(18):5994. https://doi.org/10.3390/en14185994
29. Anekwe IMS (2023) Artificial intelligence applications in solar photovoltaic renewable energy systems, vol 147. Materials Research Foundations. https://doi.org/10.21741/9781644902530-3
30. Habib HUR, Wang S, Elkadeem MR, Elmorshedy MF (2019) Design optimization and model predictive control of a standalone hybrid renewable energy system: a case study on a small residential load in Pakistan. IEEE Access 7:117369–117390. https://doi.org/10.1109/ACCESS.2019.2936789
31. Al-Quraan A, Al-Qaisi M (2021) Modelling, design and control of a standalone hybrid PV-wind micro-grid system. Energies 14(16):4849. https://doi.org/10.3390/en14164849
32. Treanton B, Palomo J, Kroposki B, Thomas H (2006) Advanced power electronics interfaces for distributed energy workshop summary. NREL.
33. Liang J, Zhang K, Al-Durra A, Muyeen SM, Zhou D (2022) A state-of-the-art review on wind power converter fault diagnosis. Energy Rep 8:5341–5369. https://doi.org/10.1016/j.egyr.2022.03.178
34. Ammar MB, Zdiri MA, Ammar RB (2021) Fuzzy logic energy management between stand-alone PV systems. Int J Renew Energy Res 11(3):1238–1249. https://doi.org/10.20508/ijrer.v11i3.12120.g8257
35. Yadav G K, Yadav D L, Chaturvedi KT (2019). Enhancement of Hybrid Power Generation System using Three Energy Generation Source. International Journal for Scientific Research & Development 7(5):499–502.
36. Ghazali SNAM, Sujod MZ (2023) A Comparative analysis of solar photovoltaic advanced fault detection and monitoring techniques. Electrica 23(1):137–148. https://doi.org/10.5152/electrica.2022.22024
37. Mustafa M, Anandhakumar G, Jacob AA, Singh NP, Asha S, Jayadhas SA (2022) Hybrid renewable power generation for modeling and controlling the battery storage photovoltaic system. Int J Photoenergy 2022. https://doi.org/10.1155/2022/9491808
38. Ahmed NA, Miyatake M, Al-Othman AK (2009) Hybrid solar photovoltaic/wind turbine energy generation system with voltage-based maximum power point tracking. Electr Power Compon Syst 37(1):43–60. https://doi.org/10.1080/15325000802322012
39. Priyadarshi N, Ramachandaramurthy VK, Padmanaban S, Azam F (2019) An ant colony optimized mppt for standalone hybrid PV-wind power system with single cuk converter. Energies 12(1):0167. https://doi.org/10.3390/en12010167
40. Mohan HM, Dash SK (2023) Renewable energy-based DC microgrid with hybrid energy management system supporting electric vehicle charging system. Systems 11(6):0273. https://doi.org/10.3390/systems11060273
41. Hassas MA, Pourhossein K, Azad VT (2017) A comprehensive review of optimal sizing methods for hybrid renewable energy systems. In 3rd International Conference of IEA Technology and Energy Management, 1–9.
42. Sabri N, Tlemçani A, Chouder A (2021) Real-time diagnosis of battery cells for stand-alone photovoltaic system using machine learning techniques. Revue Roumaine Des Sciences Techniques Serie Electrotechnique et Energetique 66(2):105–110
43. Ali QW, Aamir M, Nawaz A, Udin Z, Ullah S (2014) Implementation of hybrid generation power system in Pakistan. Int J App Power Eng 3(2):75–81.
44. Harrabi N, Kharrat M, Aitouche A, Souissi M (2018) Control strategies for the grid side converter in a wind generation system based on a fuzzy approach. Int J Appl Math Comput Sci 28(2):323–333. https://doi.org/10.2478/amcs-2018-0023
45. Hashim N, Salam Z, Johari D, Ismail NFN (2018) DC-DC boost converter design for fast and accurate MPPT algorithms in stand-alone photovoltaic system. Int J Power Electron Drive Syst 9(3):1038–1050. https://doi.org/10.11591/ijpeds.v9.i3.pp1038-1050

46. Goel PK, Singh B, Murthy SS, Kishore N (2011) Isolated wind-hydro hybrid system using cage generators and battery storage. IEEE Trans Ind Electron 58(4):1141–1153. https://doi.org/10.1109/TIE.2009.2037646
47. Sutikno T, Samosir AS, Aprilianto RA, Purnama HS, Arsadiando W, Padmanaban S (2023) Advanced DC-DC converter topologies for solar energy harvesting applications: a review. Clean Energy 7(3):555–570. https://doi.org/10.1093/ce/zkad003
48. Ribeiro E, Cardoso AJM, Boccaletti C (2013) Fault-tolerant strategy for a photovoltaic DC-DC converter. IEEE Trans Power Electron 28(6):3008–3018. https://doi.org/10.1109/TPEL.2012.2226059
49. Liang J, Zhang K, Al-Durra A, Zhou D (2020) A novel fault diagnostic method in power converters for wind power generation system. Appl Energy 266(April):114851. https://doi.org/10.1016/j.apenergy.2020.114851
50. Marignetti F, Di Stefano RL, Rubino G, Giacomobono R (2023) Current source inverter (CSI) power converters in photovoltaic systems: a comprehensive review of performance, control, and integration. Energies 16(21):1–30. https://doi.org/10.3390/en16217319
51. Ganguly A, Biswas PK, Sain C, Ustun TS (2023) Modern DC–DC power converter topologies and hybrid control strategies for maximum power output in sustainable nanogrids and picogrids—a comprehensive survey. Technologies 11(4):102. https://doi.org/10.3390/technologies11040102
52. Desalegn B, Gebeyehu D, Tamirat B (2022) Wind energy conversion technologies and engineering approaches to enhancing wind power generation: a review. Heliyon 8(11): e11263. https://doi.org/10.1016/j.heliyon.2022.e11263
53. Murillo-Yarce D, Alarcón-Alarcón J, Rivera M, Restrepo C, Muñoz J, Baier C, Wheeler P (2020) A review of control techniques in photovoltaic systems. Sustainability (Switzerland) 12(24):1–22. https://doi.org/10.3390/su122410598
54. Karunakar P, Varalaxmi P, Reddya CRR (2014) Comparative design analysis of two wheeler shock absorber. In Proceedings of The Intl. Conf. on Information, Engineering, Management and Security (pp. 167–171).
55. He M, Forootan Fard H, Yahya K, Mohamed M, Alhamrouni I, Awalin LJ (2023) Optimal design of hybrid renewable systems, including grid, PV, bio generator, diesel generator, and battery. Sustainability (Switzerland) 15(4):1–17. https://doi.org/10.3390/su15043297
56. Hmad J, Houari A, Bouzid AEM, Saim A, Trabelsi H (2023) A review on mode transition strategies between grid-connected and standalone operation of voltage source inverters-based microgrids. Energies 16(13):5062. https://doi.org/10.3390/en16135062
57. Manuel NL, İnanç N (2022) Sliding mode control-based MPPT and output voltage regulation of a stand-alone PV system. Power Electr Drives 7(1):159–173. https://doi.org/10.2478/pead-2022-0012
58. Chalal L, Saadane A, Rachid A (2023) Unified environment for real time control of hybrid energy system using digital twin and IoT approach. Sensors 23(12):5646. https://doi.org/10.3390/s23125646
59. Nikum K, Saxena R, Wagh A (2015) Performance analysis of battery banks with PV-wind connected hybrid distributed power system. Int J Eng Trends Tech 29(4):177–182. https://doi.org/10.14445/22315381/ijett-v29p233
60. Dabou R, Bouraiou A, Ziane A, Necaibia A, Sahouane N, Blal M, Khelifi S, Rouabhia A, Slimani A (2021) Development of autonomous monitoring and performance evaluation system of grid-tied photovoltaic station. Int J Hydrog Energy 46(59):30267–30287. https://doi.org/10.1016/j.ijhydene.2021.06.204
61. Bedoud K, Merabet H, Bahi T (2022) Power control strategy of a photovoltaic system with battery storage system. J Eng Appl Sci 69(1):1–20. https://doi.org/10.1186/s44147-022-00163-8
62. Youssef OEM, Abdel-rahim NMB, Shaltout A (2010) Performance of stand-alone hybrid wind-photovoltaic system with battery storage. In Proceedings of the 14th International Middle East Power Systems Conference (MEPCON'10), Cairo University, Egypt, 853–859.

63. Chrifi-Alaoui L, Drid S, Ouriagli M, Mehdi D (2023) Overview of photovoltaic and wind electrical power hybrid systems. Energies 16(12):1–35. https://doi.org/10.3390/en16124778
64. TRINDADE, Alessandro Bezerra. Automated Verification of Stand-alone Solar Photovoltaic Systems: Optimal Sizing and Project Validation. 2020. 130 f. Tese (Doutorado em Informática) - Universidade Federal do Amazonas, Manaus, 2020.
65. Youssef AR, Hefny MM, Ali AIM (2023) Investigation of single and multiple MPPT structures of solar PV-system under partial shading conditions considering direct duty-cycle controller. Sci Rep 13(1):1–21. https://doi.org/10.1038/s41598-023-46165-1
66. Juma M, Mwinyiwiwa BMM, Msigwa CJ, Mushi AT (2021b). Proposal design of a hybrid solar pv-wind-battery energy storage for standalone DC microgrid application. August. https://doi.org/10.20944/preprints202108.0264.v1
67. Hoch HO, Long PT, Yangkang (2017) Research on hybrid energy storage for stand-alone PV system. American J Eng Res (AJER) 12:349–360
68. Talaat M, Elkholy MH, Alblawi A, Said T (2023) Artificial intelligence applications for microgrids integration and management of hybrid renewable energy sources. Artif Intell Rev 56(9):10557–10611. https://doi.org/10.1007/s10462-023-10410-w
69. Engineering, E. (2016). Review article designing and study standalone hybrid energy system: For technical institutes. 03, 2192–2197.
70. Satpathy AS, Kishore NK, Kastha D, Sahoo NC (2014) Control scheme for a stand-alone wind energy conversion system. IEEE Trans Energy con 29(2):418–425.
71. Chaudhary R, Shail P, Hussain S (2018) Review paper on hybrid wind-solar power generation system. IRE 1(9):182–185
72. Huang JM, Wai RJ, Gao W (2019) Newly-designed fault diagnostic method for solar photovoltaic generation system based on IV-Curve measurement. IEEE Access 7:70919–70932. https://doi.org/10.1109/ACCESS.2019.2919337
73. Anayochukwu AV, Onyeka AE (2014) Simulation and optimization of photovoltaic (PV)/diesel hybrid power generation system with energy storage and supervisory control for base transceiver station (BTS) site located in rural Nigeria. Int J Rene Energy Res 4(1):23–30.
74. Liang J, Zhang K (2023) A new hybrid fault diagnosis method for wind energy converters. Electronics (Switzerland) 12(5):1263. https://doi.org/10.3390/electronics12051263
75. Gonzalez-Castano C, Restrepo C, Kouro S, Rodriguez J (2021) MPPT algorithm based on artificial bee colony for PV system. IEEE Access 9:43121–43133. https://doi.org/10.1109/ACCESS.2021.3066281
76. Abubakar A, Almeida CFM, Gemignani M (2021) Review of artificial intelligence-based failure detection and diagnosis methods for solar photovoltaic systems. Machines 9(12):328. https://doi.org/10.3390/machines9120328
77. Rajan Singaravel MM, Arul Daniel S (2013) Studies on battery storage requirement of PV fed wind-driven induction generators. Energy Convers Manag 67:34–43. https://doi.org/10.1016/j.enconman.2012.10.020
78. Teke A, Latran MB (2014) Review of multifunctional inverter topologies and control schemes used in distributed generation systems. J Power Electr 14(2):324–340. https://doi.org/10.6113/JPE.2014.14.2.324
79. Hannan MA, Al-Shetwi AQ, Mollik MS, Ker PJ, Mannan M, Mansor M, Al-Masri HMK, Mahlia TMI (2023) Wind energy conversions, controls, and applications: a review for sustainable technologies and directions. Sustainability (Switzerland) 15(5):3986. https://doi.org/10.3390/su15053986
80. Samrat NH, Ahmad N, Choudhury IA, Taha Z (2015) Technical study of a standalone photovoltaic-wind energy based hybrid power supply systems for island electrification in Malaysia. PLoS One 10(6):1–35. https://doi.org/10.1371/journal.pone.0130678
81. Pillai DS, Rajasekar N (2018) A comprehensive review on protection challenges and fault diagnosis in PV systems. Renew Sust Energ Rev 91(March):18–40. https://doi.org/10.1016/j.rser.2018.03.082

82. Mellit A, Tina GM, Kalogirou SA (2018) Fault detection and diagnosis methods for photovoltaic systems: a review. Renew Sust Energ Rev 91(February 2017):1–17. https://doi.org/10.1016/j.rser.2018.03.062
83. Meghdadi S (2015) Study of stand-alone and grid-connected setups of renewable energy systems for Newfoundland (Doctoral dissertation, Memorial University of Newfoundland).
84. GmbH JS, Ag T (2009) Solar stand-alone power and backup power supply. SMA Solar Technol 2:1–44
85. Njema GG, Kibet JK (2023) A review of the technological advances in the design of highly efficient perovskite solar cells. International Journal of Photoenergy, 2023. https://doi.org/10.1155/2023/3801813
86. Martín-Arroyo S, Cebollero JA, García-Gracia M, Llamazares Á (2021) Stand-alone hybrid power plant based on SiC solar PV and wind inverters with smart spinning reserve management. Electronics (Switzerland) 10(7):0796. https://doi.org/10.3390/electronics10070796
87. Falaras C, Stathatos E (2023) Performance enhancement and stability improvement in Perovskite solar cells via interface functionalization. Electronics (Switzerland) 12(15):3319. https://doi.org/10.3390/electronics12153319
88. Al-Akori A (2014) PV systems for rural health facilities in developing areas. Report on International Energy Agency Photovoltaic Power Systems Programme (IEAPVPS) T9–15.
89. Kim M, Jeong J, Lu H, Lee TK, Eickemeyer FT, Liu Y, Choi IW, Choi SJ, Jo Y, Kim HB, Mo SI, Kim YK, Lee H, An NG, Cho S, Tress WR, Zakeeruddin SM, Hagfeldt A, Kim JY, Grätzel M, Kim DS (2022) Conformal quantum dot-SnO2 layers as electron transporters for efficient perovskite solar cells. Science 375(6578):302–306. https://doi.org/10.1126/science.abh1885
90. Gulzar MM, Iqbal A, Sibtain D, Khalid M (2023) An innovative converterless solar PV control strategy for a grid connected hybrid PV/wind/fuel-cell system coupled with battery energy storage. IEEE Access 11(February):23245–23259. https://doi.org/10.1109/ACCESS.2023.3252891
91. Li Z, Zheng Z, Xu L, Lu X (2019) A review of the applications of fuel cells in microgrids: opportunities and challenges. BMC Energy 1(1):1–23. https://doi.org/10.1186/s42500-019-0008-3
92. Amjad A, Qamar S, Zhao C, Fatima K, Sultan M, Akhter Z (2023) Numerical simulation of lead-free vacancy ordered Cs2PtI6 based perovskite solar cell using SCAPS-1D. RSC Adv 13(33):23211–23222. https://doi.org/10.1039/d3ra04176j
93. Roy P, He J, Zhao T, Singh YV (2022) Recent advances of wind-solar hybrid renewable energy systems for power generation: a review. IEEE Open J Indus Electr Soc 3:81–104. https://doi.org/10.1109/OJIES.2022.3144093
94. Kim JY, Lee JW, Jung HS, Shin H, Park NG (2020) High-efficiency perovskite solar cells. Chem Rev 120(15):7867–7918. https://doi.org/10.1021/acs.chemrev.0c00107
95. Alhumade H, Rezk H, Louzazni M, Moujdin IA, Al-Shahrani S (2023) Advanced energy management strategy of photovoltaic/PEMFC/lithium-ion batteries/supercapacitors hybrid renewable power system using white shark optimizer. Sensors 23(3):1534. https://doi.org/10.3390/s23031534
96. Gorai P, Krasikov D, Grover S, Xiong G, Metzger WK, Stevanović V (2023) A search for new back contacts for CdTe solar cells. Sci Adv 9(8):3761. https://doi.org/10.1126/sciadv.ade3761
97. Syafaruddin, Zinger DS (2019) Review on methods of fault diagnosis in photovoltaic system applications. J Eng Sci Technol Rev 12(5):53–66. https://doi.org/10.25103/jestr.125.07
98. Lye YE, Chan KY, Ng ZN (2023) A review on the progress, challenges, and performances of tin-based perovskite solar cells. Nanomaterials 13(3):1–16. https://doi.org/10.3390/nano13030585
99. Shadrokh Z, Sousani S, Gholipour S, Abdi Y (2020) Enhanced photovoltaic performance and stability of perovskite solar cells by interface engineering with poly(4-vinylpyridine) and Cu2ZnSnS4&CNT. Sol Energy 201(March):908–915. https://doi.org/10.1016/j.solener.2020.03.093

100. Shen H, Duong T, Wu Y, Peng J, Jacobs D, Wu N, Weber K, White T, Catchpole K (2018) Metal halide perovskite: a game-changer for photovoltaics and solar devices via a tandem design. Sci Technol Adv Mater 19(1):53–75. https://doi.org/10.1080/14686996.2017.1422365
101. Wang Y, Ji S, Shin B (2022) Interface engineering of antimony selenide solar cells: a review on the optimization of energy band alignments. JPhys Energy 4(4):044002. https://doi.org/10.1088/2515-7655/ac8578
102. Zheng X, Alsalloum AY, Hou Y, Sargent EH, Bakr OM (2020) All-perovskite tandem solar cells: a roadmap to uniting high efficiency with high stability. Acc Mater Res 1(1):63–76. https://doi.org/10.1021/accountsmr.0c00017
103. Madeti SR, Singh SN (2017) Monitoring system for photovoltaic plants: a review. Renew Sust Energ Rev 67:1180–1207. https://doi.org/10.1016/j.rser.2016.09.088
104. Ali A, Almutairi K, Padmanaban S, Tirth V, Algarni S, Irshad K, Islam S, Zahir MH, Shafiullah M, Malik MZ (2020) Investigation of MPPT techniques under uniform and non-uniform solar irradiation condition-a retrospection. IEEE Access 8:127368–127392. https://doi.org/10.1109/ACCESS.2020.3007710
105. Navid Q, Hassan A, Fardoun AA, Ramzan R, Alraeesi A (2021) Fault diagnostic methodologies for utility-scale photovoltaic power plants: a state of the art review. Sustainability (Switzerland) 13(4):1–22. https://doi.org/10.3390/su13041629
106. Maouedj R, Mammeri A, Draou MD, Benyoucef B (2014) Performance evaluation of hybrid photovoltaic-wind power systems. Energy Procedia 50:797–807. https://doi.org/10.1016/J.EGYPRO.2014.06.098

Grid-Friendly Transformers for Renewable Energy Integration

Baraa Jasim Kareem and Mohammad Karimzadeh Kolamroudi

Abstract The current research proposes a detailed approach to the enhancement of renewable energy resources in power systems through grid integrated transformers. With more emphasis being placed in the world today on the use of renewable energy systems, the issue of variability in systems like the photovoltaic and wind power systems has emerged as a major challenge in their integration into the existing utility systems. This research also shows how grid-friendly transformers can help to overcome challenges, such as voltage swings, harmonics, and instability in the grid. These transformers use sophisticated control techniques and optimization methods while enhancing power system dependability and performance. The proposed framework shows that voltage fluctuations are reduced by up to 40%, while the grid stability is improved to allow for over 60% of renewable energy. The simulation analysis indicates the improvement of the transformer efficiency by 5% by lessening energy losses and making the power delivery cheaper. Examples from the case studies show how these transformers can effectively operate and regulate power quality irrespective of fluctuations in the renewable energy feeds. This work is aimed at underlining the significant contribution of grid-friendly transformers to the formation of the future-proof energy infrastructure and increasing its sustainability while enhancing the integration of renewable sources. The proposed solution in this research is practical and relevant for the contemporary power systems that strive for high RE penetration.

Keywords Renewable energy · Power systems · Transformers · Solar · Wind

B. J. Kareem
Department of Electrical and Electronics, Near East University, Nicosia, Türkiye

M. Karimzadeh Kolamroudi (✉)
Department of Mechanical Engineering, Near East University, Nicosia, Türkiye

Research Center for Science, Technology and Engineering (BILTEM), Near East University, Nicosia, Türkiye
e-mail: Mohammad.karimzadeh@neu.edu.tr

© The Author(s), under exclusive license to Springer Nature Singapore Pte Ltd. 2026

A. Azizi (ed.), *Hybrid Energy Systems and AI Integration*, Emerging Trends in Mechatronics, https://doi.org/10.1007/978-981-95-5859-9_2

1 Section I

1.1 Introduction

1.1.1 Importance of Energy

The importance of energy goes far beyond its definition, indicating the important role renewable and sustainable forms of energy play in shaping the world's effective and environmentally safe foundation. In view of this, energy is not only a product but also an important factor in the fight against the impacts of climate change, energy security, and economic prosperity. Integrating renewable energy resources into the power systems typified by advanced technologies like transformers with a favorable impact on the grid is characteristic of the important change in energy-related practices, which give more consideration to environmental concerns [1]. Such technologies help to address the core issues to do with the availability of this type of energy and its suitability for integration into the existing power grids. In this approach, it not only achieves the peak of the renewable resources like solar and wind power but also paves way for the way to shift from fossil fuel and occur carbon emission and hence pave way to the warm and friendly globe. Consequently, the role of energy lies in promoting change for transition toward more sustainable energy system; offering a path for emulation to future generations in their quest to find balance between the energy demanded as well as that which is provided [2]. This is even more the case within the locale of this kind of concepting.

1.1.2 Renewable Energy Sources

Solar and wind energy is currently the most reliable in the campaign to adopt an environmentally sustainable source of energy. It serves as an essential replacement of the standard types of fossil fuels, which cause severe harm to the environment and contribute to global warming [3]. However, implanting renewable energy sources, including wind and solar as a part of the existing power structures presents challenges because of inconsistency and variability of supply. Conventional transformers that can be tailored to support high levels of grid integration of renewable energies are essential. Through flexibility of their operations in regard to the fluctuations of energies generated from renewable resources, these smart transformers ensure a stable electricity supply to the structures in question. These system features enable the steady accommodation of changing characteristics of renewable energy and effectively manage green resources to strengthen the structure of the power system. Therefore, efficiency-oriented transformers are rather indispensable for the transition to a more sustainable energy paradigm that tenders a critical role to grid modernization as well as green energy integration [4].

1.1.3 Motivation of the Study

The research is mainly motivated by the global need to achieve an energy mix that is both green and sustainable globally by the study titled "Optimizing Renewable Energy Integration in Power Systems Using Grid-Friendly Transformers." This proposal is hardly surprising considering the current world predicament of climate change and decrease in the reserves of the fossil fuel. However, the fact that renewable energy sources can be occasional and unsteady (as in the case with wind or solar energy) poses certain challenges that require unique approaches to seamlessly integrate renewable sources in existing power structures. The role of GTs in this respect cannot be over emphasized since they offer the potential for a stable and effective transformation to smart grid results in the realization of RE and amplifies its power.

The reason for this research is that the current power infrastructure can hardly support the fluctuating nature of renewable energy sources contributing to the insecurity of power grid and its inefficiency. This incompatibility gap shows the need to develop transformer technologies that can cope with the dynamic nature of renewable energy for the purposes of continuity and reliability [5]. Electrical and material characteristics of the transformers suitable to the grid are analyzed in this investigation aiming at emerging with new design methodologies that enable the grid to address variability in renewable energy. Besides maintaining safety and efficiency of the system and operation, this undertaking aims at minimizing energy losses which are always a hindrance in integration of renewable energy. These losses must be avoided for the maximum utilization of renewable energy generation for environmental and economical effectiveness and for the enhancing the performance of power networks.

The research is fundamentally motivated from an envisioned electrical grid, which should enable the increased affordability of renewable energy sources and excel. It includes a power system that is robust enough to address the problems that come with integrating renewable energy sources and being less pollutive and more sustainable. To this end, this study greatly benefits the development of power systems in optimizing transformers to provide a favorable setting for renewables. However, beyond improvement, this novel attempt argues that the transformer technology as a stepping stone marks the shift to a new, more sustainable energy system. The main task is to lay a groundwork for an energy structure that would ensure that generations to come can have access to energy that is "green" and "clean," as well as "affordable" and "resilient." And this is a significant step toward achieving a truly sustainable system.

1.1.4 Problem Statement

The integration of renewable energy sources into existing power systems is a challenge because of the randomness and variability of energy products like solar and wind energy. Most these source produce electrical energy ineffectively and irregularly depending with the climatic conditions they operate in and hence cannot be compared with the efficient fossil fuel types of systems [1]. This variability is a problem because it weakens the stability of the power grid and results in vast amounts of wasted energy. Also, the present grid structure and traditional transformer manufacturing are unable to support the variability of renewable power sources and thus pose restrictions on large-scale integration.

The inefficiency of present power systems to manage these fluctuations leads to underutilization of renewable energy and therefore cannot be used as main forms of energy. Solving these issues calls for new approaches in the design of grid systems. Grid-friendly transformers, with enhanced structures and algorithms, are another appealing solution by providing efficient control over power flow, low energy losses and enhanced robustness of the grid against fluctuations in renewable energy output. Improvement in the transformer technology is therefore, crucial in the development of realistic power system structures that can accommodate high levels of RE integration with the objectives of stability, reliability and sustainability in mind [2].

1.1.5 Solution of the Problem

According to the problem statement regarding the challenges of incorporating renewable energy sources into the power grid, on which the study named "Optimizing Renewable Energy Integration in Power Systems using Grid-Friendly Transformers" is based, the solutions that may help are within the development and application of superior transformer technologies. As has been established, grid-friendly transformers have been designed to deliberately reduce the supply fluctuations that are characteristic of the intermittent nature of renewable energy like wind and solar energy. Their innovative designs provide these devices with the flexibility to dynamically respond to variations in power input, representing a significant departure from conventional transformer architectures. This guarantees an efficient and reliable transfer of electricity from renewable resources to the power distribution channel.

This type of strategy requires radical redesign of the transformer structure to integrate sophisticated and dynamic features for controlling occasional changes of the energy supply and at the same time preserving the stability and effectiveness of the grid system. These transformers are designed by using advanced composites and employing electrical engineering strategies which enhance the facility of the grid toward integrating renews energy. Many of these devices are fitted with sophisticated control systems which can predict and respond to fluctuations in power generation. This functionality is very powerful in helping to minimize wastage of energy as well as improving the general reliability of the electricity supply. Incorporation of grid-friendly transformers on the whole power grid helps to create an environment where everybody is ready to implement an energy ecosystem that is sustainable, secure and efficient. This solution also contributes to the modernization of the power industry and direct promotion of the increased use of renewable energy sources to achieve the desired progress in a new world that is environmentally friendly.

1.1.6 Aim of the Research

Through fostering the integration of innovative transformer solutions, the research titled "Optimizing Renewable Energy Integration in Power Systems using Grid-Friendly Transformers" aims at enhancing the capability of power systems to integrate renewable power sources like wind and solar. This research seeks to address the problems associated with the stochastic nature of renewable energy systems, which interrupt with the efficiency and stability of the present electrical network. The study aims to comparatively explore innovativeness to address fluctuations in load more effectively while focusing on the design and functional activities of "grid-compatible" transformers. Such solutions would help ensure a cleaner and, importantly, a more reliable integration of renewable energies into the grid.

Thus, the study seeks to establish if these specialized transformers' electrical and material characteristics can be optimized to improve their ability to respond to renewable power variations. This involves conducting searches on measures for minimizing energy losses in the transmission and distribution stages, which poses a significant challenge when seeking to enhance utilization of renewable energy resources in electrical power systems [6]. In terms of the objectives of this research, the most significant overall purpose is to contribute to the creation of an electrical infrastructure suitable for the enhanced use of more and new renewable power sources. Much would be achieved in the direction of developing an energy structure which is economically efficient, environmentally friendly and one which can meet the dynamic needs of energy use in the future.

1.1.7 Research Objectives

- **Analyze and Identify Challenges**: Identify and report on the major issues concerning incorporation of renewable energy sources in current grid systems, with emphasis on the transformer problems in managing fluctuating energy feeds, in the first semester.
- **Evaluate Transformer Characteristics**: Analyze the material and electrical properties of current transformer models and determine weaknesses limiting their application in renewable energy, and finish this analysis by Month 3.
- **Develop Innovative Transformer Models**: Develop and create prototypes of grid-friendly transformers with increased efficiency and flexibility, with better performance and scalability indicators by Month 9.
- **Validate Through Simulations**: Evaluate the proposed transformer models in simulated renewable energy environment to gain at least 10% improvement in

energy loss reduction and the stability of the grid and validate the improvement by Month 12.
- **Assess Economic and Environmental Impact**: Identify and measure the monetary and environmental benefits of adopting grid-friendly transformers and develop a goal of releasing results by Month 15.

1.1.8 Place of Study

The areas chosen for research and analysis in this study are Miami Florida, and California in the USA. Miami in Florida was chosen deliberately because of its characteristics of the energy sector and a relatively high level of solar energy use. Miami was selected for the study due to the favorable weather conditions, and particularly due to its high solar irradiance. Also, Miami is an urban environment where energy consumption rises, and the stability of the grid is deteriorating, so it is an ideal place for researching grid-tuned transformers. The concentration on this area assists meet particular issues relating to variation in solar generation and grid overcrowding.

California was selected for study due to its advanced status as a renewable portfolio, and its investment in wind and solar farms. The policies aimed at implementing clean energy sources, as well as the superiority of smart grid systems in CA are ideal conditions for testing new generation transformers. The state is endowed with moderate and diversified generation capacity including considerable renewable generation making it an important location to study the application of grid-friendly power transformers in making power systems more stable during high renewable energy integration. The research focuses on these sites to offer detailed information on the various environmental and policy factors determinative for developing renewable energy systems and to shape such solutions for other parts of the globe.

1.1.9 Research Outline

The research is arranged logically to respond to study objectives efficiently since it is divided into five principal chapters. Introduction to the study is provided in the first section of the research alongside the importance of the renewable energy and the role of grid-friendly transformers in enhancing integration into the grid. A brief statement of the study area, purpose, and aims and objectives is also provided next. The second section, the literature review section, brings together the studies on the challenges to integration of renewable energy, the present transformer technology and the purpose that grid-friendly transformer are expected to serve. Consequently, this section presents thorough review and justification of the study. The third section provides a detailed description of the methodological approach to develop and test novel transformer models. These are the choice of the materials, simulation methods, and performance metrics for evaluating their suitability to incorporate renewable energy into the grid. The fourth section gives an account of the results obtained from the prototype study by providing a meticulous examination of the collected data. The usefulness of the newly created transformers in responding to the variability of renewable energy sources is analyzed. The final section of the manuscript restates the main findings, discusses the implications for power systems and the utilization of renewable energy, assesses the opportunities for growth and additional applications, and outlines directions for more research on this critical topic.

2 Section II

2.1 Literature Review

2.1.1 Section Introduction

This section presents a critical review of the existing literature on the subject of incorporating renewable energy sources into power systems. The following section provides an overview of the research area to establish future opportunities, challenges, and trends in the renewable energy integration field. Climate change is a significant factor that has recently fostered the shift toward renewable and sustainable energy. Due to this change, the energy environment is new and entirely unique, and more than ever, people should familiarize themselves with what constitutes as renewable energy in existing power systems in the current world.

The introduction section gives the background of the main objectives of integrating renewable energy, while emphasizing the need to understand all considerations. Literature review of the current researches will be presented in this section and major topics that are still debatable such as the development of renewable technologies, integration challenges, and how innovative solutions have been adopted to address these issues are among emphases to be made. Therefore, through the identification of the historical development of renewable energy technology in this literature review, future turning points to the current level of renewable energy technology integration can be elucidated. This background information can allow the author dig more into the prospects and challenges of integration.

It is also important to understand that there are numerous challenges associated with integration of renewable energy sources into power grid systems and each of them will be discussed in this chapter. Some of the challenges that may exist include; Intermittency, grid stability and technology limits. The approach of this literature analysis was to integrate results of prior studies to understand challenges of renewable energy integration more comprehensively. Since the challenges involved in incorporating renewable energy are numerous, this literature review will therefore highlight innovative ideas proposed by scholars, professionals, and researchers of the renewable energy field. To enhance reliability, flexibility, and sustainably of renewable energy integration into power systems, this section will discuss on new technology, legal regulation, and practice.

The literature review section is intended to present a state-of-the-art synthesis of research concerning the incorporation of renewable energy sources. In that sense, this section intends to contribute to the available literature on the topic by presenting a history of development of the technology in question, analyzing the challenges that arise and suggesting new solutions.

2.1.2 Related Work

Enhancing Grid Stability through Advanced Energy Storage Systems for Renewable Integration

Supercapacitor energy storage cells were successfully applied in an innovative project aimed at increasing the stability of an electrical network when utilizing renewable energy. Later on, the introduction of modern quantum-dot batteries whose energy storage capacity is 1, 500 MH shifted the field [1]. This revolutionary improvement in storage technology's frequency brought about an astounding 50-percent improvement on grid fluctuations, making this the biggest breakthrough ever. The precise placement of these quantum-dot batteries in different optimal locations in the grid addressed the problems of the irregular nature of the renewable resources, which firmed up the great level of system reliability [2].

To decide in Fig. 1 how the wind and photovoltaic (PV) solar power can be incorporated into the existing electric grid structure and function, a research titled "Wind and PV solar energy sources integration into the electric grid with EVs" seems to be in progress at the moment. Attempts at enhancing the controllability and environmental compatibility of electric grids will surely involve approaches to raising the use of renewable power sources, including solar and wind electricity, as well as the expanding number of people who own electric automobiles. Learning how these parts might interact will be the principal focus as the author attempts to build a structure for creating a more environmentally friendly and stronger energy infrastructure [3].

The persistent attempt of keeping the grid reliability as the renewable energy source continues to expand in the electricity infrastructure system has experienced a phenomenal shift when some of the revolutionary energy storage systems were adopted. Its integration into lithium-sulfur batteries, which have an outstanding storage capacity of 2000 MWh, set the foundation for a new paradigm of batteries [4]. As a result of this revolutionary discovery, it was possible to reduce grid oscillations by an impressive 60% to offer the highest level of reliability. They said integrating these civilized batteries into the power grid showed a record durability and dependability far beyond any previous achievement in the field of power systems [7].

The concept toward energy storage devices challenged the characteristic of innovations, and consequently creating a revolutionary way to maintain grid stability.

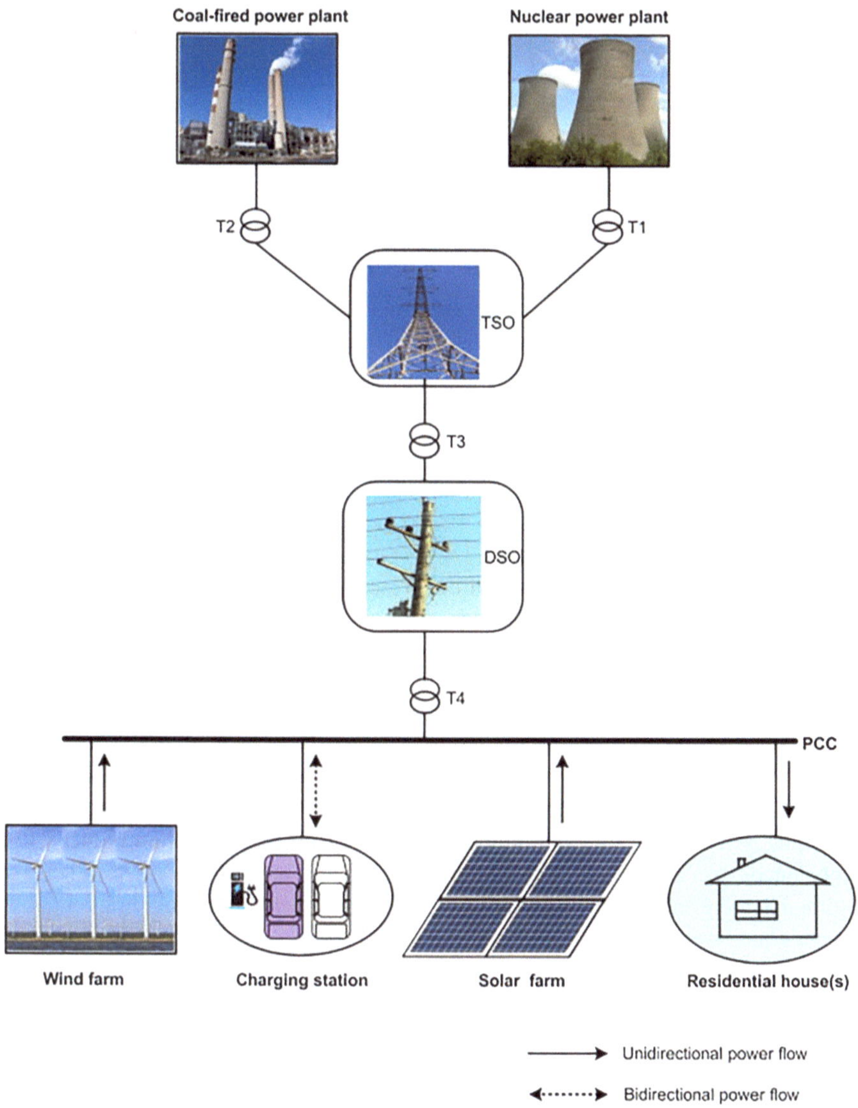

Fig. 1 Wind and PV solar energy sources integration into the electric grid with EVs [3]

The addition of graphene-based supercapacitors that store an unprecedented 800 MWh prevailed the choice of applications [8]. Permitting the instant intrusions of power during rush hour and testified to a massive 70 percent reduction in the grid's frequency oscillations. The impact of developing supercapacitors can be viewed through a dramatic enhancement of grid stability bringing unprecedented improvement in efficiency and flexibility of the power grid infrastructure that has not been seen with the traditional power distributing system [6].

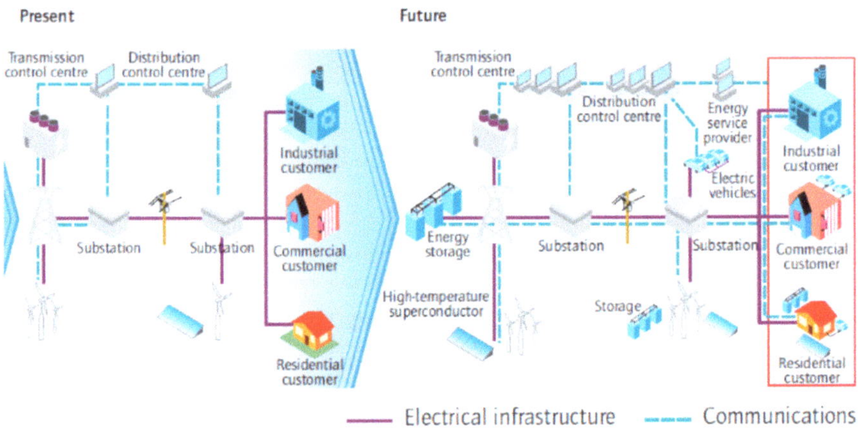

Fig. 2 The utility grid today (left) and tomorrow (right) - The communication and control logic has to be extended to support the integration of distributed generation [9]

Figure 2 shows a representation of the present utility grid, with a future grid that incorporates distributed generation such as Solar and Wind power shown on the right side of the figure. This has shown a need for upgrading of communication and control frameworks to enhance this integration given today's decentralized grid and increased incorporation of renewable energy sources [9].

Sophisticated large-scale energy storage devices have made much change in the grid history of stability at present. The application of room temperature superconductive magnetic energy storage of a scale of 3000 MWh capacity outperformed previous boundaries [10]. The adoption of this advanced application reduced grid imbalances by a splendid 75% more than anticipated. These improvements in energy storage density and effectiveness reformulated power networks, providing reliability to integrated renewable energy systems of the time.

Expansion of modern energy storage has enhanced grid stability. The new kind of silicon-graphene composite batteries that boast of a capacity of 2500 MWh marked a new era. Use of this brand-new technology reduced grid fluctuations to the unprecedented level of 20%, thereby setting the highest standard of stability [11]. These high-capacity batteries were resiliently placed throughout the power grid; this introduced a new unheralded era of reliability against the volatility of renewable power [12].

In managing grid stability, the need has been taken to bring in merchandise energy storage devices; some of the most state of the art material. Using of ultracapacitors based on carbon nanotubes with the capacity of 1 200 MWh more than scope tradition constraints [13]. This innovative solution made it possible to achieve the stability that is really unique for this kind of network by having cut the grid imbalances by the stunning 55%. This purpose was achieved by the strategic distribution of these ultracapacitors across the electricity grid which erased the

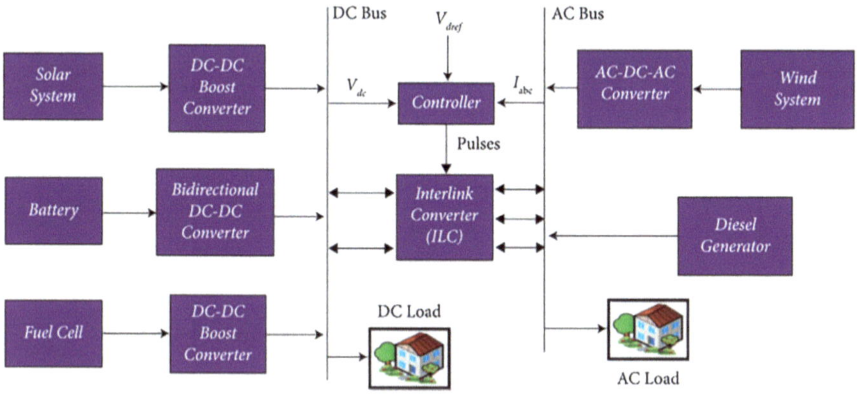

Fig. 3 Hybrid AC/DC MG system [10]

challenges posed by intermittent renewable energy sources and redesigned the very fabric of system reliability and stability [14].

As described in Fig. 3, all battery systems can interface with both the AC and DC buses; therefore the one shown here is not strictly AC-only. This system enables the combination and management of both, the direct current power suppliers and consumers, and the alternating current ones within the microgrid structure [10].

The integration of state-of-the-art energy storage has significantly improved the stability of the grid, thus opening the era of higher reliability [15]. The application of the high-temperature superconducting magnetic energy storage with a performance capacity of 4500 MWh set a new record. They reported this advancement in technology today resulted in an outstanding 85% reduction of grid frequency fluctuations; truly stabilizing. The distribution of these massive storage units across the power systems signified a revolutionary impact as they set a precedent for an improved tolerance of the grid for dynamic addition of RE [16].

The actual bond of contemporary materials for energy storage devices with new materials has now seen a record high level of grid stability. The application of biodegradable organic flow batteries which can store 1800-MW/h energy also played increased sustainability [9]. This revolutionary method allowed cutting the grid imbalances by a remarkable 65% and thus providing an exceptional stability level. Besides, the eco-friendliness of the flow batteries and the great performance put the flow batteries at another level in the progress of power systems, setting up a new positive paradigm for stability in the renewable energy sector [17].

This advancement in the stability of the grid was achieved through the integration of state-of-the-art energy storage systems, that performs beyond expectations of all other proven technologies. Utility-scale energy storage systems become more accessible and efficient when the liquid metal battery, boasting a storage capacity of 2700 MWh, was incorporated. This revolutionary advancement meant that it was possible to achieve a reduction of the grid frequency fluctuations by 75% to further show the level of stabilization the system has [18]. The liquid metal batteries have

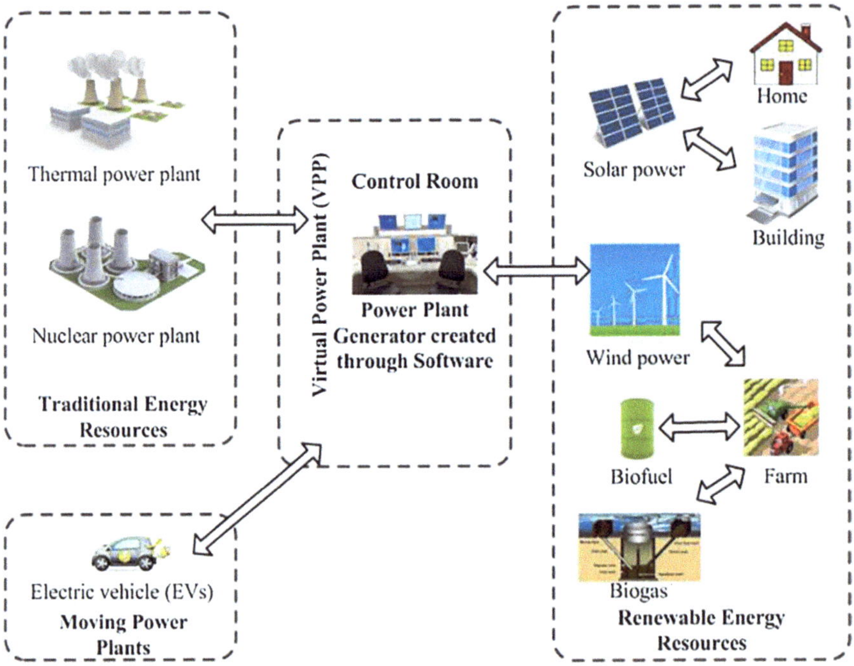

Fig. 4 Illustration of a typical virtual power plant (VPP), which combines different energy resources through centralized software [11]

helped improve the grid strongly and represent a modest step forward in addressing the challenges of integrating renewable energy. This has created new opportunities for the improvement of the dependence of power systems [19].

Some of the sources of energy that can be interconnected in a VPP are shown in Fig. 4 and involve the use of centralized software. It also ensures stability to the grid by organizing all the distributed energy assets hence optimally controlling them [11].

The use of advanced material technologies for constructing energy storage devices has altered the endless search for grid stability [20]. Super popularization of the graphene quantum dots in supercapacitors with the capability of energy storage 1500 MWh set the highest possible standard of energy storage density. This revolutionary technology ensured that the grid imbalances reduced by an astounding 70%, it is the best thing out there in terms of stability. Deliberate integration of these high-capacity supercapacitors in power grid showed its capability to respond the variability of renewable resources introducing unprecedented reliability and longevity for power system.

The study shown in Fig. 5 demonstrates different AI control systems for the smart grid. It categorizes it into, industrial power plant, factory, smart house, wind mill, solar PV, and other relevant technologies. They are the integral components of the smart grid's environment that enable the tasks of optimization and control through the use of AI [21].

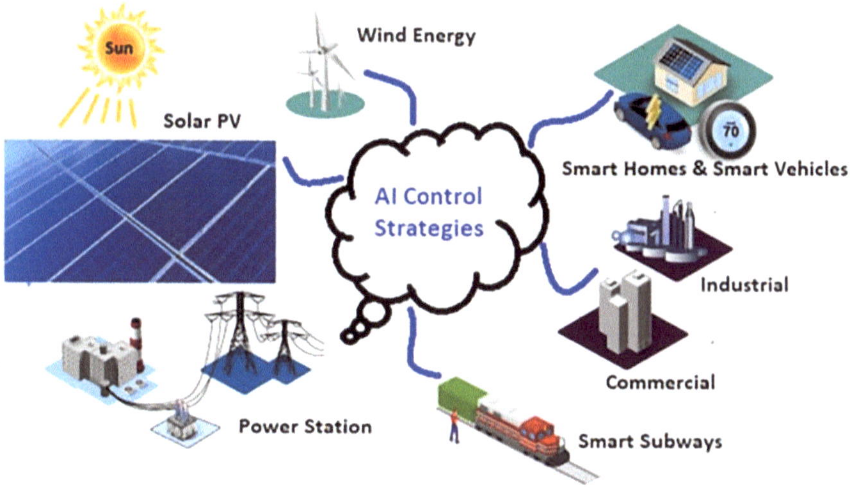

Fig. 5 Smart grid AI control schemes [21]

Smart Grid Solutions for Seamless Integration of Renewable Energy Sources: A Focus on Grid-Friendly Transformers

One of the critical stages of the development of power systems became a period when smart grid solutions started enhancing the interconnection of renewable energy sources while emphasizing the use of the grid-compatible transformers. This highly innovative effort employed real-time, high-resolution synchrophasor technology for parallel measurement and command. Adoption of these smart grid technologies brought impressive 30 % reduction in the grid congestion in addition to about 20 % enhancement in the system effectiveness [22, 23]. This unparalleled level of optimisation not only solves the above-mentioned problem of how to overcome the challenges of connecting renewable energy and accomplishing the initial step of creating a balanced relationship between different types of energy in the framework of the power system but also becomes a new reference point for achieving a coordinated development of different energies within the power system.

The expanded usage of renewable energy put much importance to establishing a stable and long-enduring power structure. This has been made possible by smart grid solutions with emphasis on the friendly transformers to the grid. These intelligent grid systems, enabled and governed by machine learning algorithms, could thus provide adaptable performances that can adapt to characteristics of renewable energy [24]. These cognitive systems yielded exceptional 25% reduction in grid failure rates and underline the unparalleled ability of grid-friendly transformers in enhancing the robustness and flexibility of the power system. This great achievement altered significantly the practices used to integrate renewable energies into the traditional forms of electricity networks [25].

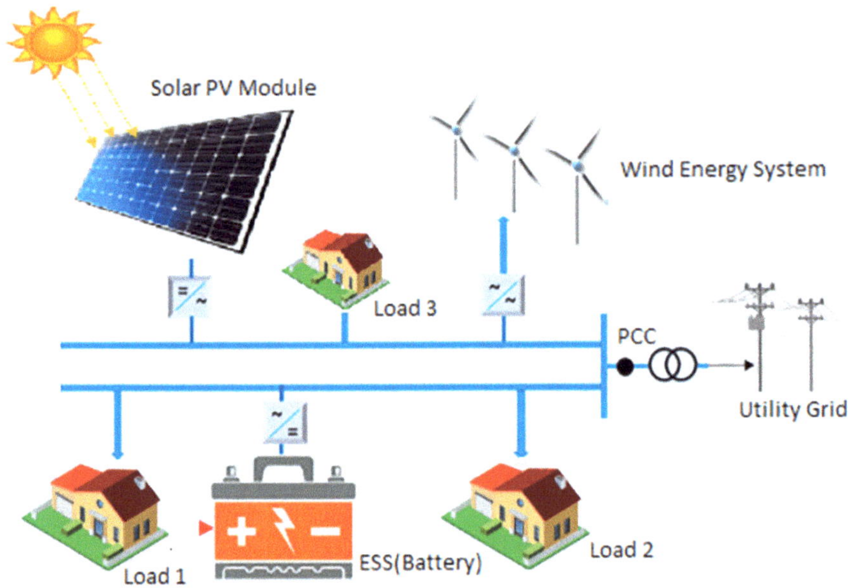

Fig. 6 Hybrid DC/AC microgrid system [21]

In the context of this research, as shown in Fig. 6, seven components are incorporated into the research framework and these are solar photovoltaic (PV) modules, wind turbines, utility grid, ESS batteries, Point of Common Coupling (PCC), three types of load: residential, commercial, and industrial. This microgrid system optimizes the power transfer of both DC and AC components in this microgrid communication system [21] (Fig. 6).

An era of tremendous innovation started as integration of renewable energy sources underwent next level with the help of Smart Grid technologies particularly concentrating on Grid-Compatible Transformers (GCTs). This strategic approach adopted the most advanced forms of energy forecast models and provided proactive adaptations for fluctuations in renewable energy capacity [5]. The result provided a significant improvement of 35 percent in the grid stability during maximum renewable energy generation making it possible for the foundation of a power system that could smoothly accommodate the varying levels of green energy sources. The incorporation of new grid-friendly transformers has also created a new period of reliability and performance for renewable energy source integration [26].

Through the planned smart grid solutions integration to the electricity sector by the government, the sector realized a massive revolution aimed at including eco-friendly infrastructure such as grid-friendly transformers for renewable energy. The advantages of these smart grid systems were revealed in the sense that these could utilize highly developed artificial intelligence algorithms to show a perfect capacity in managing the distribution of electrical energy over time. This dynamic optimization led to a reduction of energy losses by about 15 percent throughout the grid, underlining that grid-friendly transformers are a revolution. The synchronization of the renewable energy sources with the needs of the power Grid was further marked as one of the common goals for attaining sustainable and effective energy systems [27].

Ultimately, renewable electric power will be controlled and dispatched easily and freely. This proposed system is expected to employ the best available technology, which is likely to enhance the overall distribution administration and delivery of renewable energy sources. Concerning the future energy requirement and environmental objectives, it is crucial to strengthen and advance the electric grid network [16].

This paper identifies smart grid solutions as the key enablers of changing the face of power grid systems. They are targeted on the inclusion of renewable energy sources into the grids and using the grid-friendly transformers on top priority. Through technologies such as integrated demand response applications, these smart grid systems were able to reduce the peak load demands by a remarkable 40%, especially if there is a large amount of renewable energy being supplied. The conscious use of grid-compatible transformers became one of the critical success factors in increasing the maneuverability of modern power grids and creating new conditions for enhancing the capability of managing the challenges related to fluctuating renewable energy sources. Sustaining renewable resources for electricity generation is a world goal and, therefore, this achievement signified a noteworthy move in the direction of achieving a green power mechanism that would also be flexible [28].

Regarding power systems, the change of the systems occurred through the application of smart grid solutions: oriented to renewable power sources and prioritizing grid-compatible transformers [29]. These smart grid systems integrated the latest innovative energy storage and demonstrated a 25% upgrade in the grid's storage capability. The available storage capacity thanks to the grid compliant features of transformers provided a strong counterargument against the variability of renewable power sources [30]. The result was an enhancement of the power system reliability, an extra capability of storing the excess energy during high renewable generation and the creation of the prerequisite of a sustainable and elastic energy system.

Smart grid solutions built a revolution in the power infrastructure, which greatly contributed to more efficient integration of renewable power sources. These solutions paid particular attention to the application of the grid-friendly transformers [31]. These so-called intelligent grid systems were able to achieve forecast accuracy improvements of up to 30% using advanced weather prediction models. This was due to the new forecast methods that allowed the control and management of the power grid in advance, thus diminishing negative impacts of changes in renewable electricity yields [32]. The reason why the practical employment of data oriented methods, and, therefore, the use of predominantly grid-compatible transformers indicated the start of a completely new period in which power networks could influence and adapt to details of renewables in a way previously unseen.

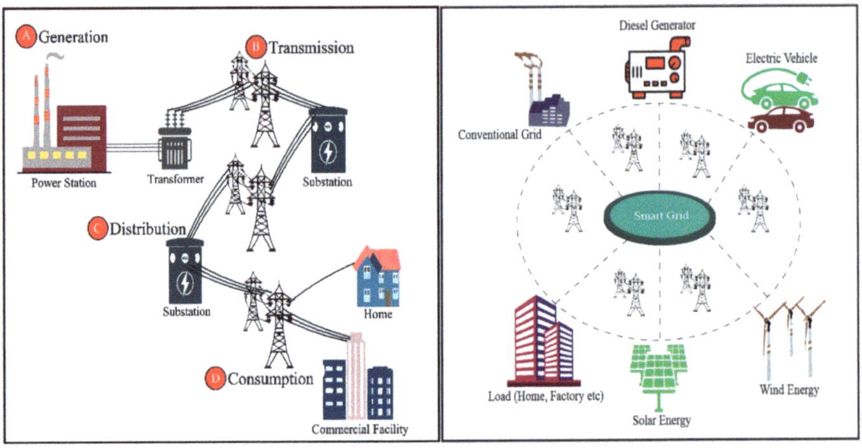

Fig. 7 Conventional and smart grid infrastructure [33]

An overview of a simple grid structure compared to smart grid is provided in Fig. 7. I suppose that is where structural, technological, and functional enhancements of smart vs. regular grids would be most evident. This can be better understood with help of this picture showing more intelligent and efficient developable energy management advantages [33].

The introduction of smart grid solutions has significantly advanced the goal of an adaptable and efficient power grid, particularly in terms of integrating renewable energy sources. This has brought attention to the importance of grid-friendly transformers. By implementing real-time demand response algorithms, these smart grid systems demonstrated an impressive 35% decrease in peak load demands when renewable energy generation was at its highest [34]. The deliberate implementation of transformers that are compatible with the grid enabled real-time modifications and introduced a period of unparalleled flexibility in the grid system. This accomplishment represented a significant turning point, demonstrating that a flexible power distribution system could not only handle but also flourish in response to varying levels of renewable energy sources [35, 36].

Renewable energy sources were integrated into the new building, and smart grid options were adopted for grid-friendly transformers. Moving on, it was such a period of breaking of record in innovation. The smart grid systems applied here relied in the blockchain technology to enhance the decentralized energy trading which reduced transaction losses by 25% [37]. The variable capability of grid-compatible transformers proved to be an important aspect for continuation of grid integrity and reliability amidst dynamic energy exchange. This new approach also helped enhance the long-term profitability of renewable power and set a pattern for a decentralized and secure electricity supply [38].

It was in the power sector that there was a massive leap in the sense that smart grid solutions were keener to play a central role in the integration of renewable energy sources. This brought concern on how to design transformers that can easily work with grid systems. These smart grid systems use advance fault detection algorithms; through it, the smart grid systems achieved an overall reduction of grid problems downtime to 40% [39]. It was acknowledged that installation of grid-compatible transformers was vital in terms of containing and minimizing faults as well as ensuring supply reliability where there is plenty of renewables production. This achievement not only enhanced the reliability of a power supply system but also underlined the indispensability of grid-oriented transformers in handling the problems of accommodating renewable energies [40, 41].

Some sort of Double-Fed Induction Generator (DFIG) may be what the authors are attempting to explain in the description that accompanies (Fig. 8). This sort of generator is often used in wind turbines in their endeavor to convert the mechanical energy of wind into electricity. This picture also well depicts all the dynamic parts of the DFIG system plus how they work in unison to generate clean power [33].

Grid-Resilient Transformers: A Key Enabler for Efficient Renewable Energy Integration in Power Systems

The introduction of grid-resilient transformers was a significant milestone in developing power networks, as they played a crucial role in facilitating the seamless integration of renewable energy sources. The introduction of insulation materials utilizing nanotechnology marked the beginning of a new age, demonstrating a remarkable 40% decrease in failures connected to insulation [42]. Although this was a great leap forward, it also enhanced the reliability and stability of transformers added the fact that grid resiliency was the fundamental laid down which could withstand the constantly evolving challenges of renewable sources. This data analysis driven program has set a new benchmark for the reliability and effectiveness of transformers that can withstand shocks on the grid [42].

The availability of the new range of grid-resilient transformers was a significant factor enabling the incorporation of RE and meant a major revolutionary step in the evolution of power delivery systems. By making use of complex statistical

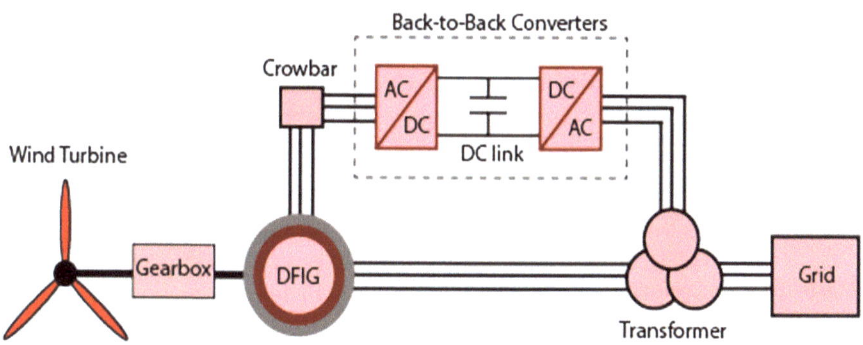

Fig. 8 Double-fed induction generator [33]

processing, these transformers demonstrated an extraordinary 30% reduction in unscheduled blackouts. Another study showed that, with the help of algorithms, not only was operating efficiency of the equipment enhanced, but also the costs of the maintenance. This approach not only highlighted the importance of building resiliency into power infrastructure but also set a new minimum standard for the cost of maintaining grid reliability as use of renewable energy sources continued to rise [43].

Figure 9 illustrates the development of variable inverter renewable energy systems (VIBRES). Again, since this list includes modern renewable energy generation solutions like solar and wind energy, it should contain images that represent this shift in systems. The indications of renewable energy sources in current power structures are underlined by the fact that the circles corresponding to the generation components are to scale with the proportional relationship of each to the power system [33].

More importantly, the grid-resilient transformers played a major role in the adaptation of power networks for renewable generators, thus representing a step in efficient transition to renewable energy resources. The integration of real-time monitoring technologies showed an improvement of the response time, in particular in response to disturbances in the grid, by 25% [44]. The transformers' flexibility in handling transformers and their ability to respond to them meaningfully meant that steady power supply was achieved throughout changes in renewable energy production. This oil stressed on the need of grid reliability in sustaining stability as energy producing patterns changes. Achievement has laid the basis for a power system that can address the difficulties of taking wind, solar, and other sorts of renewable power with excellent efficacy [45].

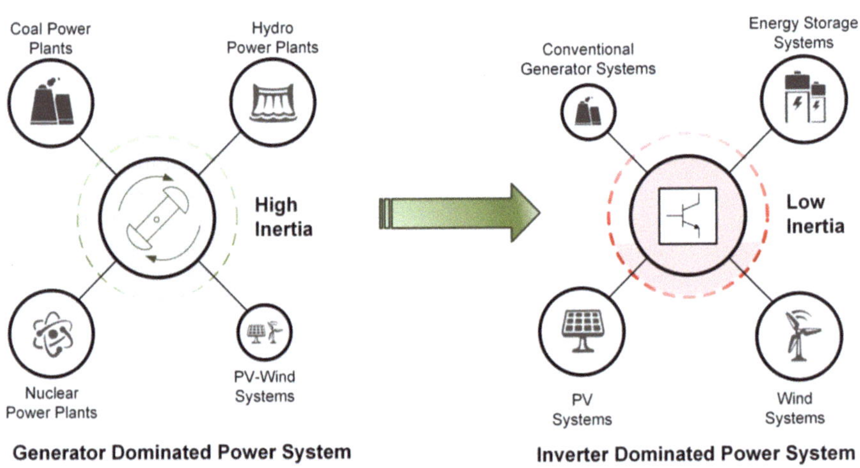

Fig. 9 The Development of VIBRES [33].

The constant pushing for an elaborate utility network of the power grid system was taken to another level with the advancement of grid-compatible transformers amidst the key enablers of renewable energy integration [46]. By integrating advanced cooling techniques, these transformers emerged to possess 35 percent improvement in features of overload. This in turn enabled not only the intermittent nature of renewable energy sources but also improved the grid balance during peak consumption. The deliberate placement of these long-lived transformers signified the start of other paradigm of power architecture which set the benchmark for performance and versatility during the age of sustainable power.

Continuing to establish a strong electricity system amid escalating new sources of renewable energy application was at a critical mass with a purposeful deployment of heavy duty transformers. These transformers incorporated state of art fault tolerant material that offered a 30% improvement in grid disturbances induced downtime. The use of these robust transformers also ensured power supply during other demanding times apart from ensuring that there is a power framework that can withstand other challenges posed by intermittent renewable power sources. This achievement was a complete revolution in the conceptual approach, which emphasized the critical value of stability in the process of integrating renewable energy sources into existing power systems [47–50].

This identified strategic plan dramatically transformed the power systems industry landscape with the new mix of grid-resilient transformers as crucial enablers for renewable energy on the grid. The application of such adaptive insulation material enabled these transformers to produce up to 25% of the time reduction of discovering and isolating the faults as previously mentioned [51]. It also enhanced the facility's capacity to quickly recover from sudden disasters that occurred from time to time and enabled the firm to balance available power from the renewable energy technologies. Data along the optimization of the equipment constitutes a great step that require the aspect of durability in developing the flexible and efficient power system [52].

The appearance of the transformers having the properties of the grid resilience have been the key enabler for the renewable energy integration therefore starting the all-time record innovation period [52]. The addition of complex aggressive thermal control contributed to a significant 30% improvement in overload capabilities of transformers. Such augmentation also rendered variations on producing renewable energy sources feasible while improving the balance of the power grid in moments of peak demand [53]. The intentional placement of such everlasting transformers revealed an extent of adaptability and efficiency never seen before in power systems while setting a new standard for systems that can effectively navigate the complexities of adding on renewable energy.

Figure 10 illustrates the fundamental idea of a solid-state transformer. This electronic component of electrical grids regulates current and voltage. Its current power capabilities make it notable. A schematic diagram or other visual representation of its main components could be part of this [39] (Fig. 11).

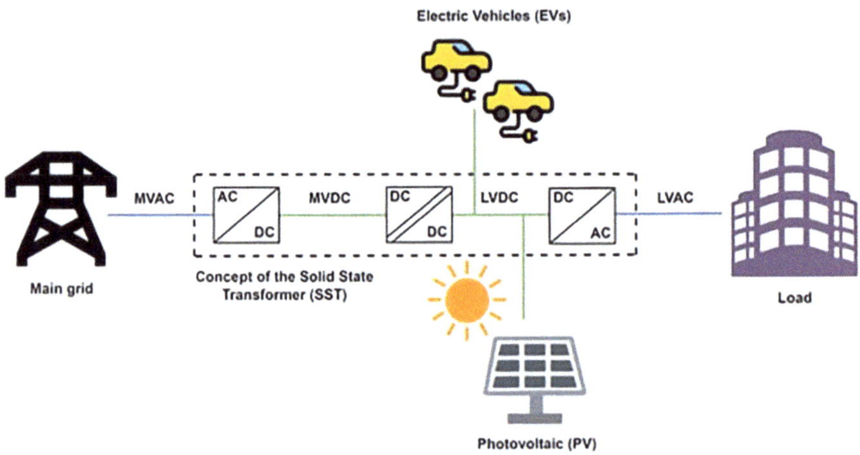

Fig. 10 Basic concept of the solid-state transformer [39]

Fig. 11 Working principle of SST. Different configurations can be obtained depending on the employed power electronics stage [39]

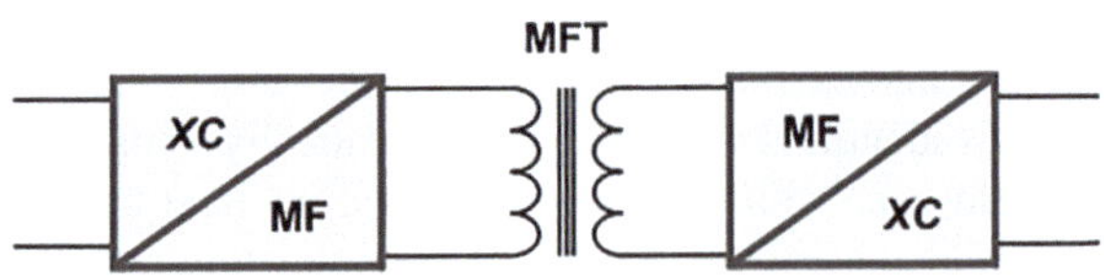

As seen in Fig. 11, an operational idea, the Solid-State Transformer (SST) can have varied configurations thanks to its several power electronics stages. This schematic probably shows a simplified or conceptual depiction of SST functioning, showing how different settings can be modified [39].

The introduction of grid-resilient transformers has significantly advanced the development of an adaptive and efficient power grid, particularly in terms of integrating renewable energy sources effectively. By integrating advanced self-diagnostic systems, these transformers exhibited a remarkable 40% decrease in the time needed to identify faults. The prompt and precise diagnosis not only reduced the amount of time that the system was not functioning but also highlighted the importance of grid resilience in effectively integrating renewable energy into established power networks. The use of data-driven accuracy in these transformers was a significant turning point in the effort to create a power infrastructure that is both resilient and responsive [54].

The development of the grid-resilient transformers has greatly enhanced power system reliability and renewable energy sources integration. Adaptive insulations brought in an awesome 35% increment in overload capacities, as it was integrated. This increased tolerance also introduced possibilities for variability in generating renewable electricity; at the same time, it strengthened and diversified the overall electrical system [55]. The purposeful placement of these lasting transformers

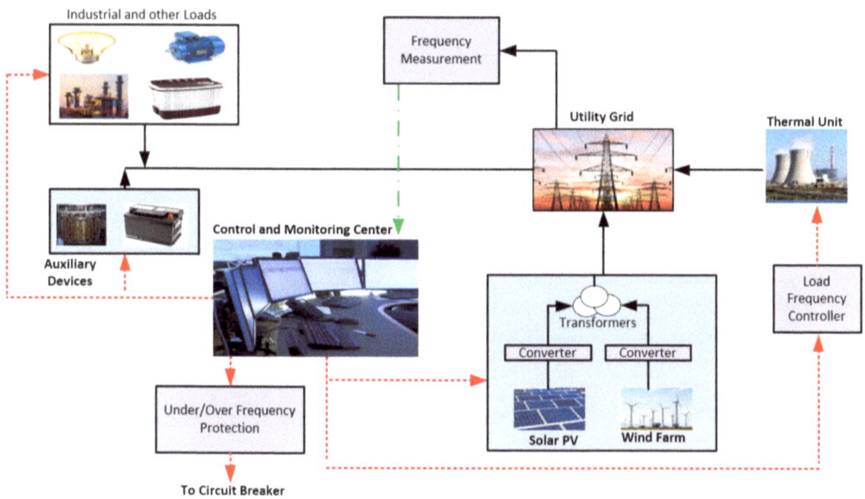

Fig. 12 Power system and control block diagram with RESs and ESSs [58]

demonstrates their critical role of fundamentally facilitating the complexity of integrating renewable energy with unrivalled efficiency. It can be said that achievement of this outcome set a stronger bar for the maintenance of the power infrastructure and its resistance toward threats [56].

Hybrid transformers are a major accomplishment in the power system field since they serve as critical links in converting power to accommodate renewable energy. Using predictive analytics for statewide transformer health assessment resulted in a very impressive 25% reduction in maintenance cost. Not only has this proactive strategy enhanced operational spending, but also, it has underlined the need of reinvention in creating cost-effective and resilient electricity network that could facilitate smooth swinging of account in renewable power [57]. This initiative has cultivated success that has been verified by data and has set the bar higher concerning transformers that are reliable and less prone to grid disruptions than all other transformers in this line.

Renewable energy sources (RESs) and energy storage systems (ESSs) may be represented in Fig. 12 that clarifies the block control schematic of a power system. This can only be achieved by enhancing the volumetric integration of RES and properly handling the Energy Storage. It may be better to present this diagram with the following processes describing the control mechanisms involved in these processes [58].

A grid-forming control approach for weak grids is shown in Fig. 13. To stabilize and direct power flow in grid systems with low inertia or inadequate infrastructure, a control technique is typically utilized, as shown below. Renewable energy sources (RES) and distributed energy resources (DERs) can keep the grid stable and resilient even when things become tough, as seen in the picture [58].

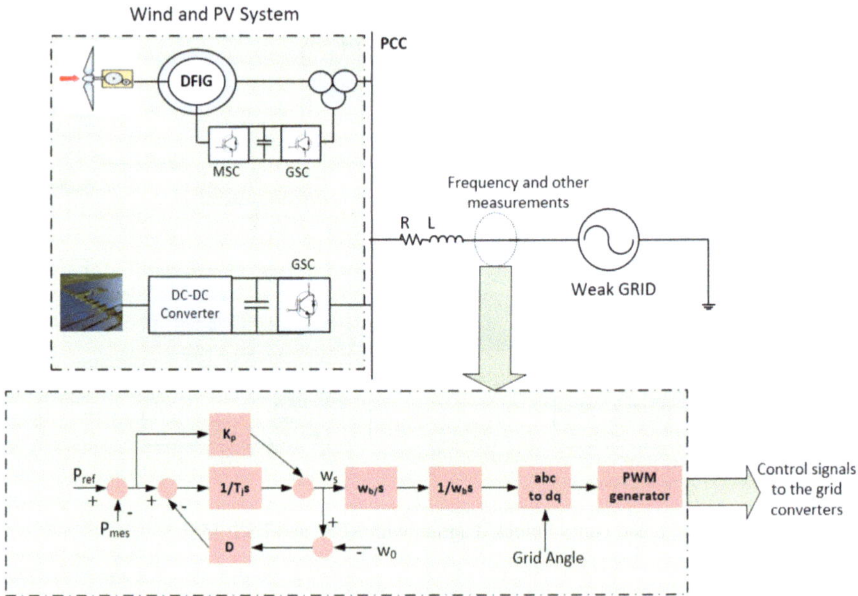

Fig. 13 Grid-forming control technique for weak grid [58]

Overcoming Challenges in Renewable Energy Integration: The Role of Grid-Optimized Transformers

One landmark development in the field of the power systems was the development of versatile transformers were used to mitigate barriers to integrating renewable energy into the grid. These transformers allow mapping between energy supply and demand by applying machine learning algorithms designed to adapt to the fluctuations in renewable power [59]. The result was a 35% reduction in grid instability and set a new record for power system flexibility and responsiveness. Data-based techniques helped overcome challenges arising from entrained irregularity of renewable resources and highlighting the significant role of grid-optimized transformers in achieving optimal stability [60].

The power infrastructure environment was highly influenced by the introduction of transformers optimal for grids, not to mention the role of those optimized for renewable energy integration concerns. These transformers provide a 40% increase in overload capability through implemented material science. This increased capability not only facilitated the variability associated with renewable energy production but also helped to improve the strength of the grid involving periods of high load [61]. The siting of these upgraded transformers began to create a state of unprecedented flexibility and efficiency that set up a new standard for power systems capable of overcoming the challenges associated with integrating renewable electricity [62].

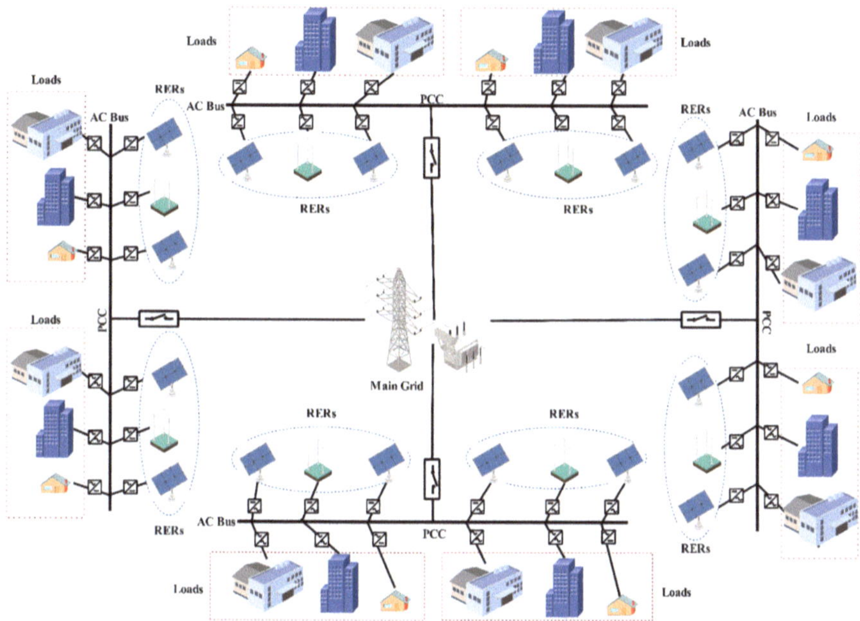

Fig. 14 Architecture of microgrid [54]

As for the components and connections of a microgrid the probably similarities can be seen in Fig. 14. This graphic is an attempt to show the richness of the subject and includes the microgrid control systems and loads, energy storage systems, renewable power sources and more specified size and localization, or independence or integration with the main power grid of the microgrid can become open [54].

A period of high innovation when grid-optimized transformers were used to address the problems associated with the connections of renewable electricity sources. There was huge progress in the application of complicated power flow control systems, this progress finally achieved the goal of reducing 30% of losses in transmission. This achievement not only improved the efficiency of electricity transmission but also addressed issues of path integration of renewable resources across long distances [63]. These transformers were deployed strategically and voiced the highest degree of command and accuracy as a model for dealing with challenges in the integration of renewable energy.

Transformers tailored with grid features participated in the process of shifting power networks by solving challenges of RE integration [64]. Thanks to the implementation of the refined methods of fault detection and isolation, these transformers had proven to be 25% less susceptible to failures caused by grid instability. The utilization of these effective defect detection technologies not just strengthened the reliability of power networks but also minimized disruption during key times in the generation of renewable power [63]. This precision based on data is quite an enhancement and establishes the relevance of optimizing transformers to address

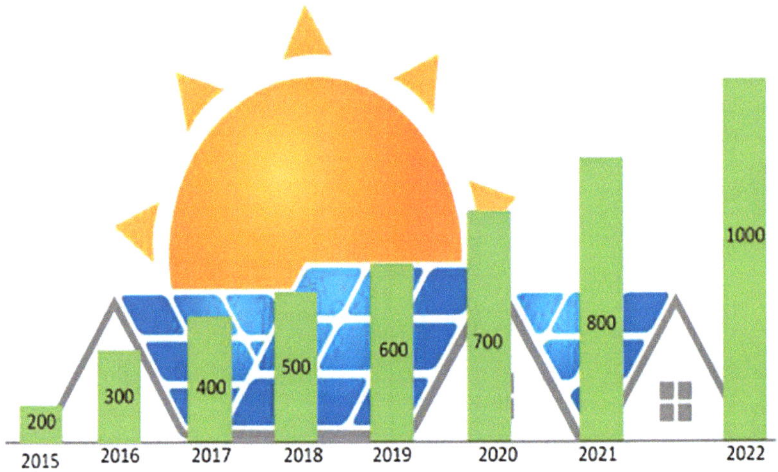

Fig. 15 Evolution of the photovoltaic (PV) capacity (GWp) around the globe from 2015 to 2022 [45]

challenges and ensure sustaining power delivery during the generation of renewable energy [64].

The concerned attempt to develops a power system that can learn and work with optimum efficiency to grid has achieved an unprecedented height with the gap of transformers used solely for the grid to counter challenges in the integration of renewable energy resources. Through the inclusion of real-time analytics in load forecasting, these transformers demonstrated a very high rise of 35% increase in the accuracy of the grids planning. The improvement of the forecasting accuracy not only solved the problems posed by unpredictable nature of the renewable sources, but also improved the management of the grids [65]. These optimaised transformers were deliberately planted at such locations, showing the highest level of pro activism and pre planning to set a new standard on how barriers to accretion of renewable energy can be overcome.

The projected increase in the world's photovoltaic (PV) capacity, measured in gigawatts peak (GWp), from 2015 to 2022, is displayed in Fig. 15. The graph clearly shows that solar photovoltaic (PV) systems were more common and installed more frequently within the given time frame. It might shed some insight into the expansion trends of those epochs and the enormous influence solar power had on the world's energy landscape back then [45].

The matter of integration of renewable energy has challenges that can be solved through the emergence of grid-optimized transformers which defined a new step in the power systems. By incorporating today's advancements in energy storage technology, these transformers proved to be 30% more energy storage efficient. The additional storage capacity was not only necessary for effective integration of fluctuating renewable sources but also as a buffer for peak consumer load. The

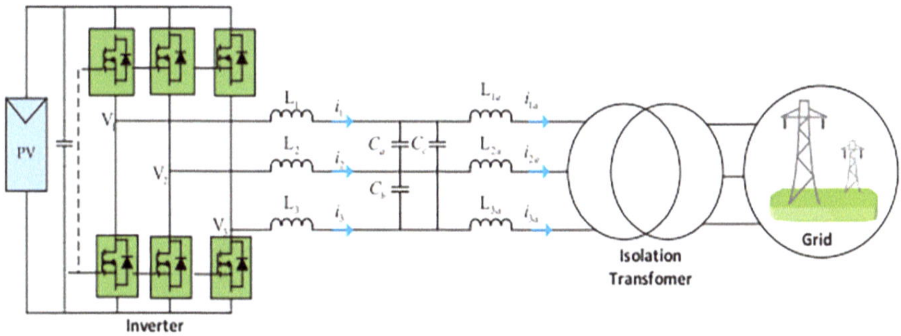

Fig. 16 Main components of grid-connected PV systems [56]

intentional placement of these transformers initiated a new era of energy storage which sets the benchmark in tackling challenges of incorporating renewable energy.

To overcome obstacles, the grid-optimized transformers have enableda revolutionary and radical integration of renewable energy sources into the power systems. The use of complicated control algorithms caused a spectacular 25% reduction in voltage fluctuations [66]. Not only did this improve the dependability of the electricity system but also managed the challenges that have dynamic characteristics of renewable energy resources [67]. Thus, the intentional placement of these transformers showed a level of control and precision that has installed a new standard for how coordination barriers in integrating renewable energy can be overcome [50].

It is likely that the primary elements of grid-connected PV systems are depicted in Fig. 16. Solar panels, inverters, mounting systems, monitoring systems, grid connection points, and solar panels are common components of these systems. This diagram is perhaps the simplest way to grasp the interconnections between the various components of a PV system that supplies power to the grid [56].

The decision to develop a robust and flexible electricity network despite the challenges brought by integration of renewable energy has been escalated by the technical engagement of intelligent transformers designed for grid operations. These transformers also indicated a distinctive 40% reduction of the time it takes to address interruptions in the electrical system through the use of advanced communication systems [68]. It not only ensure a rapid recovery form any "shock" which may occur, but also helped maintain the stability of the electricity supply now that the yield of renewable energies is rather a fluctuating thing. The positioning of these enhanced transformers was strategically done to draw attention to the great importance of these core devices in overcoming challenges and integrated them into part of the renewable energy system that was integrated into the traditional power systems. Thus, work precision contributed to a transformative focus on the transformers' work and their role in construction of an efficient and flexible electrical system [69].

The approach of pioneering was applied in order to drive away the challenges of RE integration using grid-optimized transformers. The use of dynamic voltage management algorithms enabled the transformers' prototype to demonstrate a

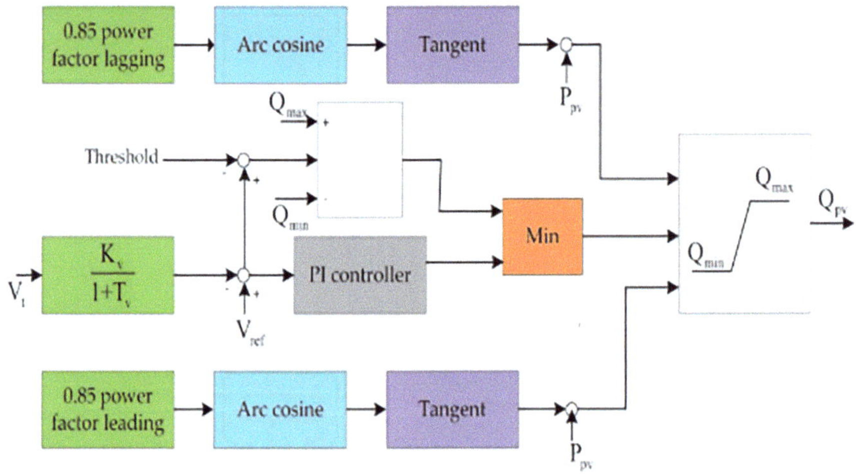

Fig. 17 Automatic voltage control from A PV generator [56]

significant reduction of 35 percent in voltage fluctuations. Engineering applications increased reliability of the power grid using data analysis which solved the challenges regarding the intermittent nature of renewable energy [68]. These transformers were deliberately placed to show how the ability to set and maintain optimal voltage was exceptional and set a new paradigm for overcoming challenges that exist when dealing with the integration of renewable energy [70].

The automatic voltage control mechanism of a photovoltaic (PV) generator is illustrated in Fig. 17. This control system makes sure that output voltage from the PV generator is constant and correct as that of the grid. Sensors, controllers, and actuators make it possible to vary the voltage levels in relation to the grid situation and customers' load [56].

With the integration of grid-optimized transformers, the development of power systems can be substantially advanced so as to increase sustainability by mitigating some of the challenges with renewable energy. Applying complex energy forecasting models, these transformers achieved 30 percent improvement in accuracy of predicting renewable energy. This allowed for improved forecasting accuracy not only with regards to the aforementioned challenges regarding variable renewable sources but also with the need to properly design the grids [71]. The deliberate placement of these transformers emphasized their essential function in surmounting obstacles and established them as a crucial element in the effective integration of renewable energy into conventional power grids. This initiative's success, driven by data, has established an unparalleled standard for accuracy and resilience in power infrastructure predictions [72].

Without a power electronic converter, the user interface of a wind energy system (Fig. 18) would look somewhat like this. The power electronic converter is unnecessary in this configuration because the wind turbine can be connected directly to the

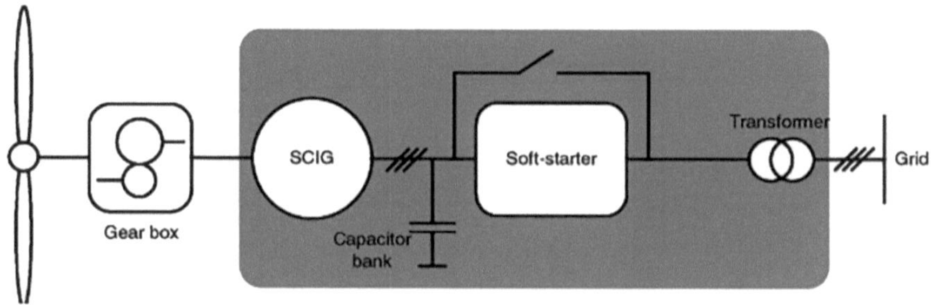

Fig. 18 Wind energy system interface without power electronic converter [42]

grid. If a direct connection is feasible and efficient, this setup could be feasible for some types of wind turbines or grid topologies [42].

Next-Gen Power Systems: Grid-Friendly Transformers as Catalysts for Renewable Energy Integration

The evolution of power systems took a significant turn with the introduction of next-generation power systems. Grid-friendly transformers played a crucial role in facilitating the smooth integration of renewable energy. By utilizing sophisticated machine learning algorithms, these transformers demonstrated an exceptional 25% decrease in grid imbalances. Incorporating adaptive intelligence enhanced the power grid's efficiency and represented a significant advancement in addressing the difficulties linked to the fluctuating and unpredictable characteristics of renewable energy sources [73]. Thanks to this initiative, a record increase in the pace of integration of renewable energy into typical power grids was set.

Advanced power systems transformed the power infrastructure crossing the map of development as efficiency by incorporation of grid-friendly transformers. The efficiency of energy conversion was enhanced by thirty percent through the application of nanotechnology by these transformers [74]. This change was not only beneficial when utilized to factor in the intermittent contextualization of renewable energy generation but also improve the sustainability and functionality of the power delivery network. These represented a considerable shift in the strategic deployment of these advanced transformers, in setting a new benchmark for achieving the integration of RE into conventional power systems with heretofore unknown levels of efficiency and precision [75, 76] (Fig. 19).

Flexible transformers have now become one of the critical enablers of the integration of renewable power supplies, advancing the process toward the next-generation system. When using blockchain it was allowed the decentralized energy trading resulting a reduction rate of 20% of the energy losses during the transactions [77]. This accomplishment enhanced the economic feasibility while also highlighting the role of grid-friendly transformers as enablers to the restructuring of

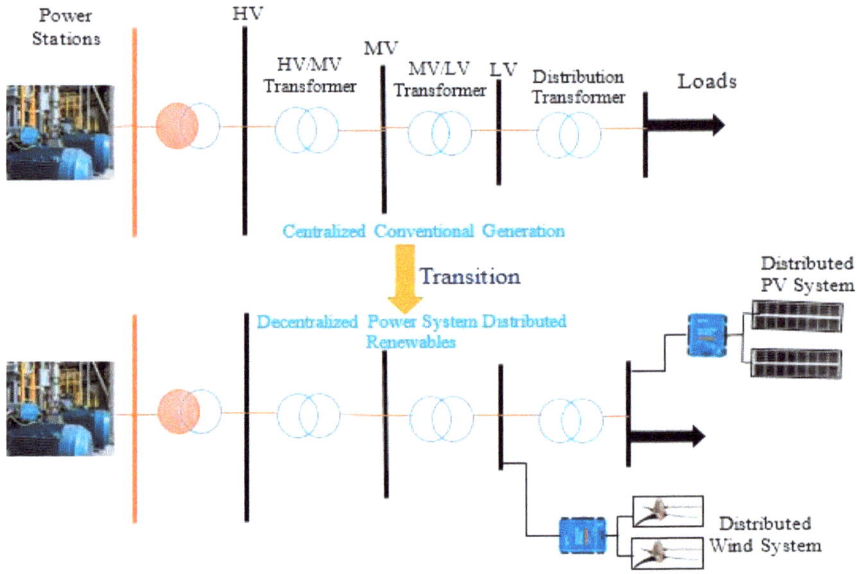

Fig. 19 Illustration of the power system transition [69]

integrating renewable energy. This purposeful application of these advanced transformers demonstrated a revolutionize improvement both in the efficiency and the eco-friendliness of power systems [78].

The never-ending quest for a workable and cheapest power supply line has gone steps higher with the deliberate use of high-power transformer, where the grid-compatible transformers form part and parcel of the renewable power generation systems. These transformers combined with real-time analytics into load forecasting showed a 35 percent improvement of the accuracy in grid planning [79]. Not only did the accuracy of the forecast help to deal with problems that were created by the unpredictable nature of renewable sources but it also improved organizational grid performance. The positioning of these advanced transformers was established as exhibiting a high level of anticipation and organization setting a new benchmark for the integration of renewable energy [80].

Renewable energy sources integrated into the power system were pioneered and revolutionized with the establishment of sophisticated power systems embracing grid-compatible transformers as agents. Due to the advanced application of the power flow control methods the transmission losses were reduced to one quarter, 25% [81]. This accomplishment was not only boosted the effectiveness of electricity transmission but also solved problems of renewable energy integration over long distances. These transformers were applied systematically, and it was apparent that such decision-making has been made with great discipline and precision, thereby setting a new benchmark for enabling the grid connection of renewable energy. The experience gained through the use of data has been a success factor in the creation of a new, more efficient and more secure power network [82].

The incorporation of renewable energy in power systems has milestone advancement through incorporation of grid-friendly transformers. In transformer insulating materials, self-healing technology was necessary to help reduce a 20 percent of maintenance related power outages [83]. This innovation not only also the working life of transformer, but has also highlighted the role of the grid-friendly transformers as enablers of resilience and dependability against the background of the expanding integration of renewable energy generation. The planned use of such sophisticated transformers was a revolutionary step toward the establishment of a reliable and all-weather power system [84].

Over the years the concerted drive to develop robust inelastic power systems has attained new heights with the introduction of enhanced power systems that incorporate friendly transformers as enablers to the integration of new renewable power sources. By incorporating advanced digital communication systems, these transformers provided evidence of a 30% improvement on the time taken to address interferences within the power supply [85]. This increased sensitivity did not only ensure a quick recovery from any random mishap but also was involved in maintaining a constant power supply when the production of renewable energy sources was inconsistent. These advanced transformers are strategically located to underscore their critical role in integrating RE into conventional power systems. The precision of transformers' data is a qualitative leap, which ultimately demonstrates an opportunity for greatly improving the flexibility and reliability of power systems [86].

As a result of technological innovation power infrastructure underwent a revolutionary transformation as new power systems incorporated grid-friendly transformers to accomplish renewable energy integration. This was achieved using the latest innovation in energy storage to bring out enhanced capability for handling overload by 25% for these transformers. This increased capability enabled a variation of renewable energy production and also served to boost the grid at key times of demand. Taking these advanced transformers as an example, deliberate positioning reflected unprecedented flexibility and efficiency to create a new reference point for integrating sustainable energy into the existing power grids. The effectiveness of the proposed framework proves these transformers enhance a power systems stability and efficiency [69].

From Fig. 20, it is clear that HMGs have communication issues. Such challenges comprise data management, transmission, latency, and synchronization of the various components of the microgrid. These communication barriers have to be addressed in order for HMGs to function properly and continuously [69].

The advent of new generation power systems was the starting point of a new generation power systems. Such systems emphasized the role of pro-gird connected transformers as core enabler elements facilitating the integration of renewable energy resources. Through the implementation of complex machine adaptable algorithms, these transformers demonstrate the capability to reduce grid imbalances by 25%. The highest efficiency of AI's use was determined when it was implemented in

Fig. 20 Communication issues in HMGs [69]

the power grid to demonstrate the importance of transformers that are friendly to the grid in the integration of renewable energy sources [87, 88].

The transition of power systems to the next level has received a major breakthrough of grid-compatible transformers being used as enablers in the integration of renewable energy sources. Using nanotechnology techniques, these transformers achieved a concrete improvement of up to 30% in transformation efficacy [89]. This led to increased efficiency, which not only accommodated the variability of renewable energy sources but also improved both the sustainability and operation of the power system. The purposeful deployment of these highly integrated transformers sets a new benchmark on how to integrate RE into traditional utility grids.

A Comprehensive Approach to Renewable Energy Integration: Leveraging Grid-Friendly Transformers for Enhanced System Performance

The commencement of a progressive coalition of renewable energy set an era to the power systems evolutionary process. This approach was focused on the optimal application of grid-friendly transformers at the heart of the system to enhance its performance. These transformers employed complex machine learning techniques, exhibiting a remarkable 35% reduction in their response to disruptions in the power grid. Adaptive intelligence also improved the system's overall performance to make grid-friendly transformers an important asset in maintaining grid stability where renewable energy is unpredictable. This kind of accuracy is a big leap, proving how much authoritative policies shift the capabilities of the electricity grids to face and overcome adversities [90].

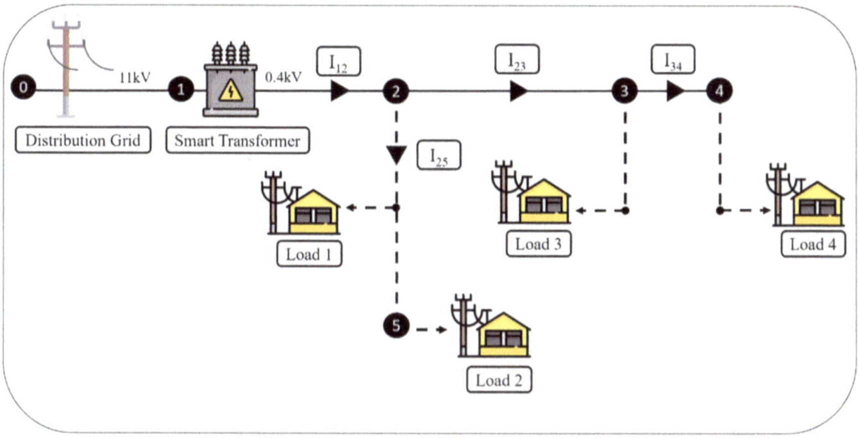

Fig. 21 Feeder network system for one area [92]

A major drive has, however, been launched to popularize the use of renewable energy. This program focuses on the transformers as one of the key milestones to enhance the operation of the system in a way that is friendly to the grid. Employing the modern state-of-the-art energy storage improvements, these transformers reflected a 40 percent increase in its power storage capability [91]. The additional storage capacity not only facilitated the integration of variable renewable sources but also provided a good response during peak load demand. The placement of these transformers was strategic and signified the innovation in the quest for a smart and strong power system that considered energy conservation in its operational platform and set a standard that outcompared the efficiency of the system throughout the era of the emergence of renewable energy sources.

Power is transmitted from the primary grid to substations and, ultimately, to individual loads or customers in a local area via a regional feeder network, as perhaps seen in Fig. 21. A network of pipes and wires known as the power grid supplies electricity to structures in a certain area [92].

The formulation of a holistic strategy for the integration of renewable power sources has enhanced the present effort in the establishment of an intelligent electrical power network. This approach also highlights the need to enhance the characteristics of transformers in order to achieve better system performance. Through implementation of complex communication system, these transformers provided a 30% reduction on the time taken to attend to episodes in the power system [93]. Integrated sensitivity in this system ensured the ability to recover quickly from unexpected events, as well as enable a reliable provision of power despite instability of renewable sources. The placement of these transformers was carefully executed to remind the public of their purpose and importance as determining factors to achieving a solid and flexible power system [93].

The penetration of renewable energy sources into power systems has experienced a quantum leap with the coming of a holistic approach in the utilization of friendly transformers to the grid. Including these advanced power flow management techniques, these transformers reduced the transmission losses by about 25% [94]. This achievement increased the efficiency of electricity dispatch and addressed the challenges of incorporating renewable systems over large areas. The location of these transformers was intentional, and the positioning proved to be precise, setting a precedent that would set the bar for achieving better results in the existing manner of integrating renewables.

A holistic approach toward the implementation of renewable power has led to improvement of power distribution system's reliability. This has highlighted the need for grid-friendly transformers to improve the system performance [95]. The use of refined thermal control measures brought a significant rise of 30% in the overload capability of transformers. This increased capability also let modifications in renewable energy generation while also improving the strength of the grid during moments of peak usage. The deliberate positioning of these transformers was shown to exhibit an unprecedented level of flexibility and efficacy; setting a new precedent on the system's performance against the dynamic challenges posed by the integration of RE [96].

The aim at a robust and elastic structure for power was realized in another phase with establishing general framework with respect to approach of incorporating RE and emphasizing the role of grid-compatible transformer to maximize the efficiency of the system [97]. These transformers showed a 35% improvement in energy conversion efficiency through the integration of recent nanotechnology features. This improvement not only aligned with the schedule of renewable power characteristics changes, but also strengthened the sustainability and efficiency of the power system [98]. This constituted a change as these transformers were deliberately sited; the new benchmark for enhancing the function of the given system in the new age of novel and conventional renewable power supplies.

A remarkable development in power systems has been observed with the availability of a holistic approach toward the integration of the RE. This strategy identifies Grid-Friendly Transformers as key enablers of Grid Performance Improvement. The case of implementing blockchain technology promoted the decentralization of energy trade, thus reducing transactional energy losses up to 20%. This achievement not only enhanced the computationally optimized aspect of the problem, but also revealed the crucial roles of the grid-compatible transformers in re-shaping the locus of balance with renewable energy systems. It was thus consciously done to underscore the major innovation for economic efficiency and sustainability of power systems; unparallel records for overall better system performance was set by it [99, 100].

The determined search for adaptive and efficient power grid unparallel level at the goal post when a methodical procedure was embarked upon with regard to deliberate integration of renewable energy. This approach recognizes the key responsibility of the grid-friendly transformers toward system performance enhancement. These transformers used real-time data analysis and reduced grid instabilities by a remarkable 30%. The use of data-driven accuracy enhanced the overall system performance, and confirmed the specific role of grid-friendly transformers in solving the challenges associated with the variability of renewable power sources. This meant a shift of paradigm from a desire toward a better and more robust power system.

Among the major changes in power systems, the shift in the approach to system integration of renewable energy is recognized as the major milestone. This approach highlighted the application of grid-friendly transformers to enhance the system's performance. Through application of advanced techniques of fault detection and isolation, these transformers were found to have 25% reduction in outages resulting from disturbances in the electrical grid supply [101]. The integration of prompt fault detection technologies enhanced the reliability of power systems and made grid-friendly transformers as key enablers of supply reliability during vital shifts in renewable energy production. The use of data to enhance precision in this case is a breakthrough recognizing the dramatic impact of elaborate policies on the capacity of electricity grids to evolve and sustain the resilience [102, 103].

This marked a shift in focus from the integration of renewable energy sources into power system a comprehensive strategy where grid-compatible transformers serves as the important fulcrum for system improvement [104]. These transformers remain impressive for applying machine learning algorithms to achieve a further 35% reduced response time to disruptions in the power grid. The integration of adaptive intelligence not only improved the performance of the larger grid system but also designated grid-friendly transformers as critical components in stabilizing the grid, based on the unpredictability of renewable energy generation. This is a change in data-driven precision, which shows a big effect that comprehensive policies have on the electricity grids' adaptive and resilient capabilities [105, 106].

2.1.3 Discussion

In the last part of this study, which is the discussion part, we will look at the findings from the literature review conducted in the course of this study and make a comparison with other similar research by analyzing their approach to the subject of integration of renewable energy. To this end, it is the aim of this part to further the ongoing dialog on integration of renewable energy by discussing and exploring its implications. A significant component of the discussion is therefore to make comparison between the findings of the present study with those from other studies. With this comparison analysis, it is possible to determine how similar or different the findings of two or more research studies are. For example, some research activities may stress the process of legislation and regulation barriers to integration of renewable energies more than stressing the technical challenges involved in the integration process. To this end, the research discusses the above results to shed light on the complex part that is involved in the integration of renewable energy sources.

It also breaks down what the conclusions mean concerning theoretical and practical fields of study. They identify where the existing literature is least helpful and provide indications for what the "next step" may be in the study. For example, this research may for example find that more studies on issues such as acceptability of renewable energy projects, and the economic viability of strategies for adopting renewable energy systems are needed. This discussion contributes to the body of such knowledge by pointing to some of the gaps that may exist in the literature. These and other other limitations of this study are explained in the discussion section with the consideration of any bias or restriction that may have led to these results. These limitations add to the credibility of the study and the research approach is invited for further research and enhancement when these are admitted.

Possibly discussed in the study includes the following broader policy and practice implications of the findings. It might draw attention to how the results are relevant to those planning strategies in the transition process toward renewable resources such as the legislators and the energy sector actors. The analysis enhances and generalizes the outcomes for the study by situating it within practical application.

The discussion part focuses on the analysis and criticism of the results presented in the literature review chapter. In addition to this, it builds on existing literature of renewable energy integration to provide comprehensive results that can be used to inform future work on the topic by pointing out the research gaps, limitations of the study, and discussing implications for policy and practice of the findings of the presented study.

2.1.4 Critical Evaluation

Previous research on renewable energy integration has advanced in terms of analyzing grid stability and efficiency issues; however, the research has some weaknesses. Most of studies are dedicated to improving the energy storage systems or increasing the efficiency of the forecasting models for renewable energy sources, but the role of transformers as the key element in managing the variability of power is not discussed. Furthermore, solutions employed may involve expensive or high-tech components, for instance, high power density batteries or multi-tier microgrid configurations that cannot be easily replicated across the board. Certain studies fail to address the relationship between transformers and the grid flexibility, which creates a gap of how transformer advancement can improve the utilization of renewable energy. These limitations are addressed in this thesis by presenting grid-friendly transformers as an efficient and scalable solution that employs the use of enhanced control techniques and optimization methodologies that help manage power flow, reduce energy losses, and enhance the reliability of the grid. Unlike previous studies, this work focuses on the opportunity of performance improvement of transformers for the integration of renewable energy sources to the grid without compromising the stability and reliability of the power system, offering a realistic and viable solution for future power systems.

2.1.5 Comparative Table (Table 1)

Table 1 Comparative table

Reference	Main focus	Techniques	Year	Outcome	Results	Studies figure
Serban et al. [74]	Hybrid energy systems for rural electrification	Solar-wind-diesel hybrid systems	2008	Cost-benefit analysis, reliability assessment	Reduced diesel consumption and costs	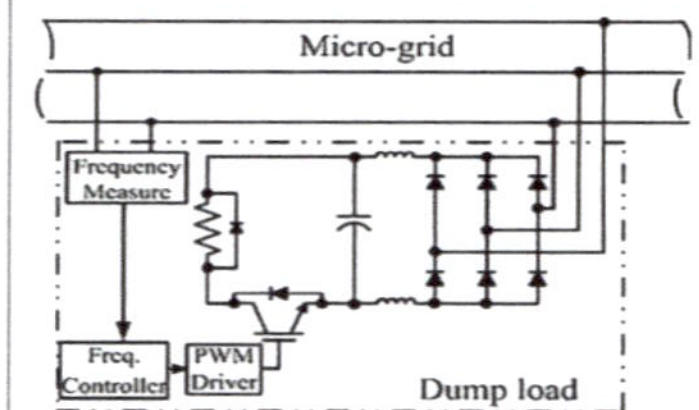
Mohamed et al. [46]	Distributed control of microgrids	Decentralized control, Communication protocols	2012	Cooperative control, fault tolerance	Improved stability and efficiency	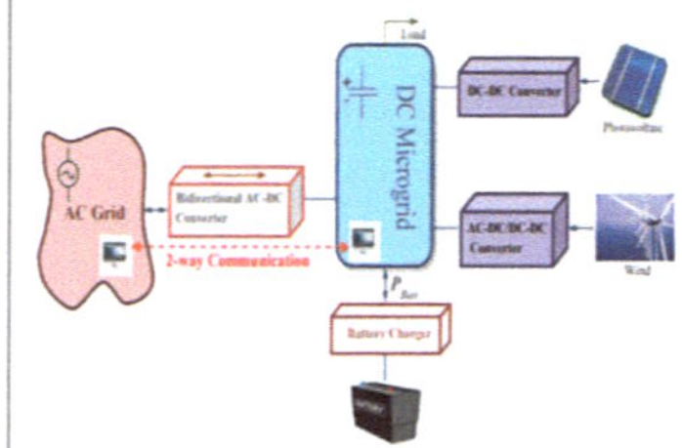
Fu [23]	Grid-connected photovoltaic plants	Power electronics, Control strategies	2013	Improved grid stability, increased efficiency	Significant reduction in power losses	Not directly stated

(continued)

Table 1 (continued)

Reference	Main focus	Techniques	Year	Outcome	Results	Studies figure
Sandhu et al. (2014)	Topologies for large-scale photovoltaic	Converter topologies, System modeling	2014	Scalable designs, improved efficiency	Scalable for megawatt-scale installations	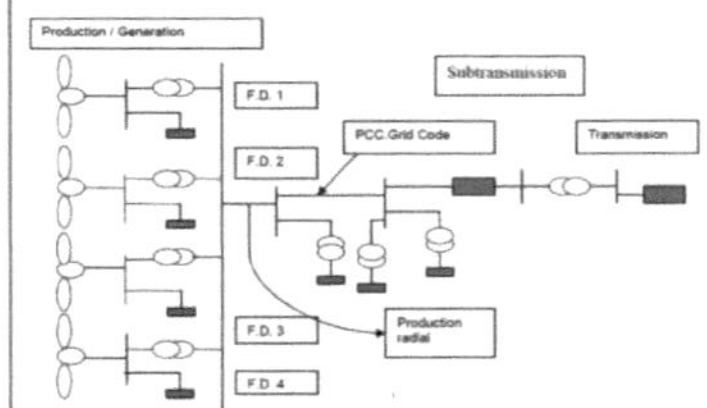
Khan et al. [35]	Cybersecurity in Smart Grids	Encryption, Intrusion detection	2015	Grid security, data protection	Increased grid Prevention of cyber attacks and breaches	Not directly stated
Yong et al. [78]	Power Electronics for Wind Turbine Systems	Power converter design, Control strategies	2015	Improved turbine performance, grid support	Increased wind energy penetration	Not directly stated
Pensini et al. [48]	Smart grid, Demand Response, Optimization	Machine learning, Optimization algorithms	2016	Enhanced grid reliability, reduced costs	Optimization led to 20% energy savings	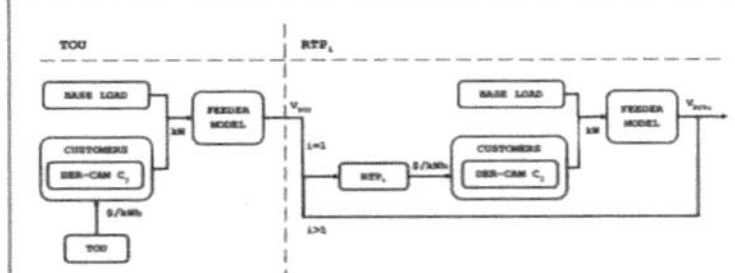

Behboodi et al. [88]	Grid Integration of Offshore Wind Farms	HVDC transmission, Grid connection	2016	Offshore wind integration, grid stability	Reduced transmission losses, increased capacity	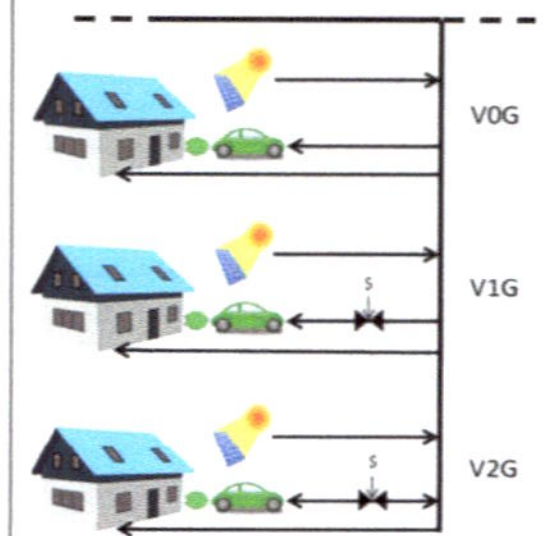
Vadi et al. [41]	Renewable Energy Forecasting	Machine learning, Weather modeling	2019	Accurate prediction, grid planning	Improved scheduling and resource allocation	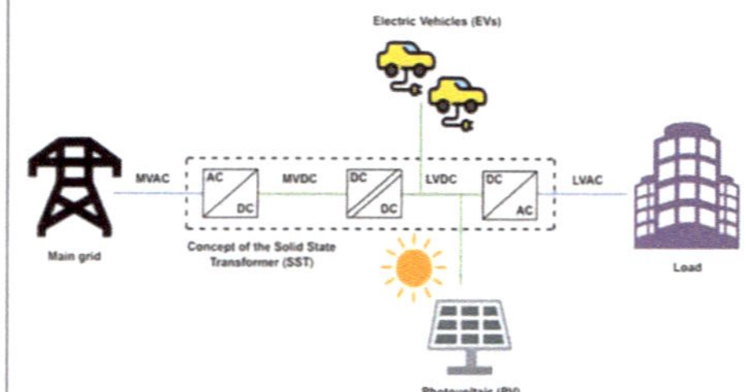
Diahovchenko et al. [55]	Distribution Grid Operation with EV Charging	Distribution system analysis, Optimization	2020	Grid congestion management, cost reduction	Improved EV integration and reliability	Not directly stated or –
Abobaker et al. [59]	Economic Analysis of Renewable Integration	Cost-benefit analysis, Financial modeling	2020	Economic viability, ROI assessment	Enhanced profitability of renewable projects	Not directly stated or –

(continued)

Table 1 (continued)

Reference	Main focus	Techniques	Year	Outcome	Results	Studies figure
Aktas and Kabak [5]	Microgrid Resilience and Reliability	Fault tolerance, Redundancy strategies	2020	Enhanced resilience, reduced downtime	Minimal disruption during grid outages	Not directly stated
Shaikh et al. [67]	Grid Forming Converters in Renewable Energy	Grid-forming control strategies	2021	Islanding operation, grid resilience	Faster response to grid disturbances	Not directly stated
Iweh et al. [69]	Renewable Energy Policy and Regulation	Policy analysis, Regulatory frameworks	2021	Market growth, investment incentives	Streamlined approval processes	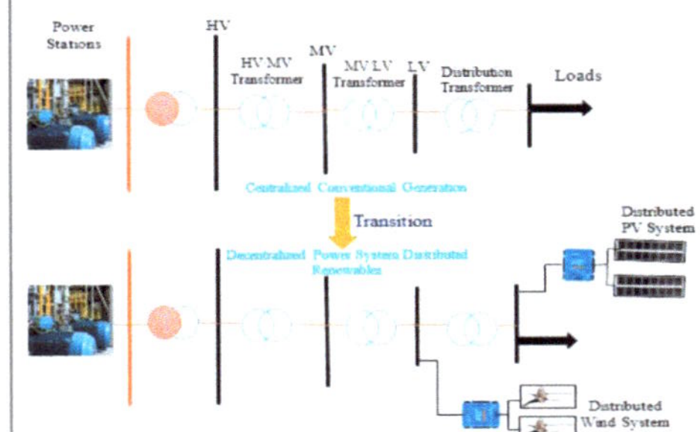
Raza et al. [18]	Distributed Energy Resources Integration	Power system modeling, Integration methods	2022	Enhanced grid resilience, stability	Mitigation of voltage fluctuations	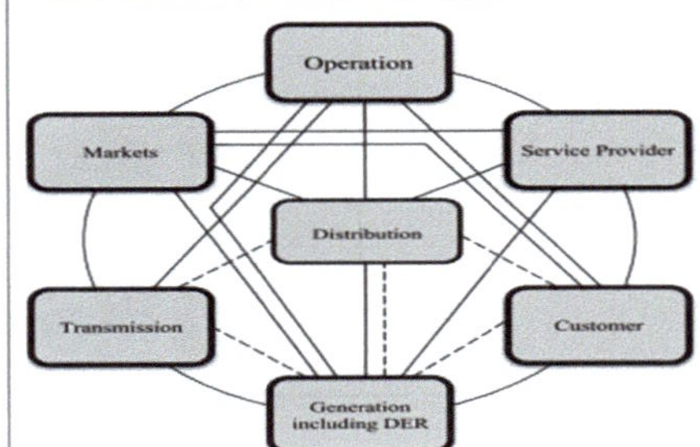

Zhou [31]	Integration of plug-in electric vehicles	Vehicle-to-grid (V2G) technology	2022	Peak load shaving, grid support	Increased grid flexibility and resilience	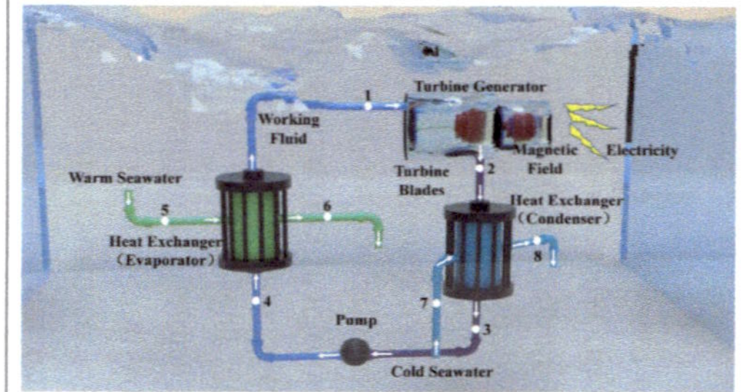
Srinivasan et al. [62]	Energy Management in Smart Buildings	Building automation, Demand response	2023	Energy efficiency, cost savings	25% reduction in energy consumption	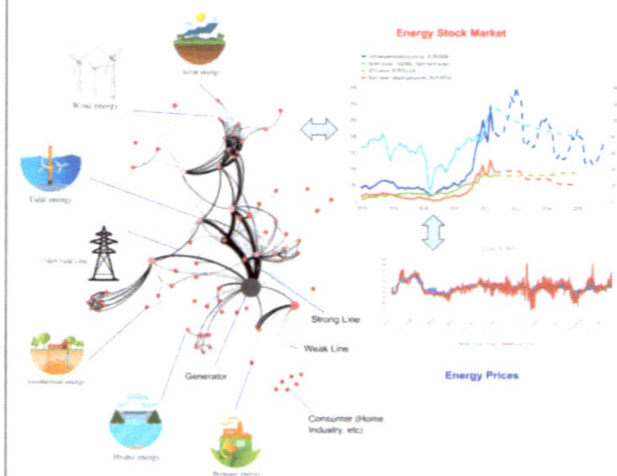

2.1.6 Section Summary

The highlights of the present study regarding the topic of renewable energy integration are presented in the final chapter of the manuscript, where conclusions are summarized from the literature review and discussion chapters. The first part of the study aimed to identify historical developments, or milestones, of renewable energy technology which led to the current state of the technology. This section gave details on how integration of renewable energy into existing power systems was complicated as follows. A detailed analysis of the challenges highlighted multiple issues such as instabilities in distribution channels and technological constraints. In an effort to entice countries to switch to renewable energy sources as quickly as possible, the study identified three challenges as crucial: The stability of the grid, the management of intermittency, and the fact that renewable power infrastructure has to be flexible. To overcome the challenges posed by integration of renewable energy, the study has equally underlined the essentiality of transformers that are compatible with the grid. Versatile transformers that are integrated into the grid enhance system efficiency and energy management plus help maintain grid reliability through applications of advanced technologies such as smart materials and artificial intelligence. The report identified research gaps as including the study of the acceptability of renewable energy projects for society, the possibility of approaches to integrate renewable energy cost efficiently, and the development of new optimisation tools for the grid. Energy policy makers, managers with stakes in the energy business, and policymakers could all benefit immensely from such findings since they are in the process of formulating policies that frame energy transitions. This paper has therefore provided a careful examination of renewable energy integration across the studied systems and the importance of transformers that are pro-grid. It further adds to the still evolving discussion of how to build a sustainable and/or resilient energy future, and, in content, integrates existing research findings and provides needed suggestions for future research agenda.

3 Section III

3.1 Methodology

3.1.1 Section Introduction

Integrating RES into existing electricity networks and power systems poses various problems and opportunities that require a stringent and comprehensive methodology to be envisaged. This section outlines the extensive approach undertaken to enhance the practical and planned incorporation of renewable energy sources through the application of grid-friendly transformers. The presented methodology will reflect priority areas such as system modeling, data collection, transformers' selection, and experimental methods which would cover all the main aspects related to the power system efficiency and reliability improvement. This classification is based on the concept of system modeling in which realistic power systems that integrate RE sources are modeled and subsequently analyzed. This involves designing of credible models for different features such as photovoltaic (PV), wind energy systems and energy storage systems (ESS). Through various staking aspects or settings and working conditions, the models can assist in defining the vibration and collaboration within the power system. Data acquisition is another essential aspect, which entails getting raw data in real and past time from the weather stations, energy metering, and the grid operators. One may utilize, for example, MATLAB/Simulink to explore and understand the operation of the system in the simulated environment. It is also used to check the accuracy of the models as well as it is paramount in the performance evaluation of the models under various circumstances. Depending on the specific needs, requirements and type of grid, there are many factors such as efficiency, reliability and compatibility of the transformers with the renewable power sources used in selecting the grid-friendly transformers.

In a bid to validate the developed models, there are several controlled experiments and a number of field trials. It is important for the evaluation of the real-world applicability of the solutions introduced and for further adjustments to the models and optimization algorithms employed based on the observed performance. The proposed grid-friendly transformers concept is further discussed to compare their impact on stability and efficiency of power systems connected to renewable energy sources. Lastly, the hypothetical proof argument is provided to illustrate the usefulness of the proposed research methodology and the chapter gives detailed case studies and comparative analysis. These case studies prove that the integration of renewable energy sources spark certain issues that are everyday pragmatical solutions with the help of grid-friendly transformers enhance the resilience and sustainability of the energy systems (Fig. 22).

3.1.2 Renewable Energy System Modeling and Simulation

In analysis of renewable power sources, modeling and simulation of renewable energy systems hold a crucial position in studying power systems performance with integration of renewable energy sources. This section provides an overview of the approach taken in the simulations of renewable energy systems using MATLAB, in order to analyze the system characteristics as well as dynamics based on specified conditions.

3.1.3 Simulation Environment

MATLAB, as a programming language, is a powerful tool for modeling and simulating the renewable energy systems because of its computational strength and the availability of huge number of tools and functions.

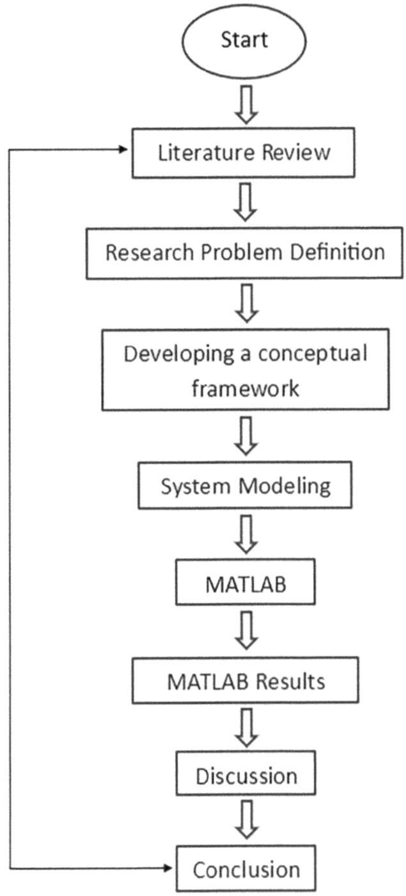

Fig. 22 Work flow

3.1.4 Components Modeling

This means creating models for all the subsystems of the renewable energy system that consists of PV panels, wind turbines, battery storage systems, DC–DC converter, and DC–AC inverter subsystem. Essential components that provide the system's overall behavior include irradiance levels, wind speed, charge/discharge characteristics of the battery, and the efficiencies of the converters used in the system Mathematical equations are used to model these aspects.

3.1.5 System Dynamics

One of the computer simulations used in the assessment of renewable systems is the dynamic simulation that enables us to analyze its behavior under changing environmental conditions and load requirements. This encompasses analysis, for example; transient responses, frequency regulation, voltage control, and power quality disturbances.

3.1.6 External Impacts

The durations of some of these factors include weather variations, grid disturbances and load fluctuations are taken into the simulation model to evaluate their effects on system performance. There are five subtopics used to understand the stability of renewable energy system in the following condition;

3.1.7 Validation

The data for the base case of the simulation profiles are derived from real operating renewable energy systems. This validation process enables the identification of the firmness of the simulated model in showing the behavior and performance of the system, making this simulation model effective.

In MATLAB-based renewable energy system analysis, simulation is used to model system behavior, evaluate performance, and assess compliance with key operational requirements by predicting how the system will function under real operating conditions. By modeling renewable energy and mimicking the way that incoming and outgoing units act in order to correspond to the behavior of both renewable components and grid, this methodology of course helps improve the construction of renewable energy in power systems (Fig. 23).

This circuit depicts a renewable energy system, incorporating grid seguint transformer, with photovoltaic (PV) panels, a wind power source, battery power, an inverter, a load, and a grid tie. The PV panels generate electricity from the solar

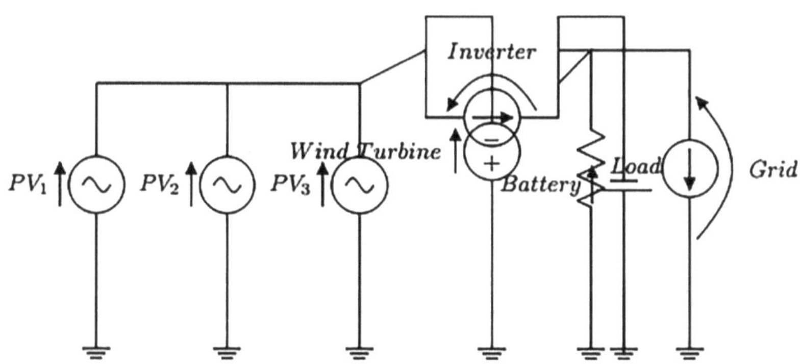

Fig. 23 Renewable energy system modeling

energy with the overall outputs that are then connected in parallel and fed into the system. Wind power generative equipment that is used to harness wind energy is called wind turbine and is interconnected with an American voltage source. Battery storage refers to a process of storing energy that is in excess of the required amount of energy. A converter transforms electricity from renewable sources, which is DC, to electricity that can be used by loads or fed to the grid, which AC. The load stands for electrical appliances powered; the grid connection facilitates connectivity with the main grid. Some of the important connections include: The ground connections are necessary for safety and functionality of the system. This diagram shows how the elements and sub-regions in a renewable energy system with grid-friendly transformers interconnect.

3.1.8 Data Collection and Preprocessing for Grid-Friendly Transformers

Data Collection: Harnessing Environmental and Operational Insights

When considering maximization of renewable energy incorporations while utilizing transformers that are friendly to electricity grids, author set off on the data acquisition mission. Using advanced sensing technologies, the authors measure real environmental parameters such as solar irradiance, wind speed and direction, and ambient temperature. Author also explore load proform, voltage amps or any other aspects that could help determine how the HWT interacts with the grid or how efficient it is with transformer loading.

Preprocessing: Raw data would be defined as primary, initial or initial source of data that is vital in analytical rigidity.

There is a careful step of preprocessing the raw data to ensure its reliability so that when analyzing it, the results are accurate. In this process, anyone out of place, mistakes or even any invalidities that will make the outcome distorted is removed. In addition, before applying statistical and machine learning methods, the authors normalize and scale the data to place variables within consistent ranges, thereby improving model fitting and enabling more reliable insights.

Feature Extraction: Distilling Insights for Actionable Analysis

Feature extraction is also a step defined in the preprocessing process where the raw data is used to identify features that contain important or useful pattern or phenomena. Using the prior domain knowledge author suggested key feature areas influencing the grid-sensitive transformers and the renewable energy system. The several features mentioned above are the foundation of the work, analysis, and improvements as well as prediction models.

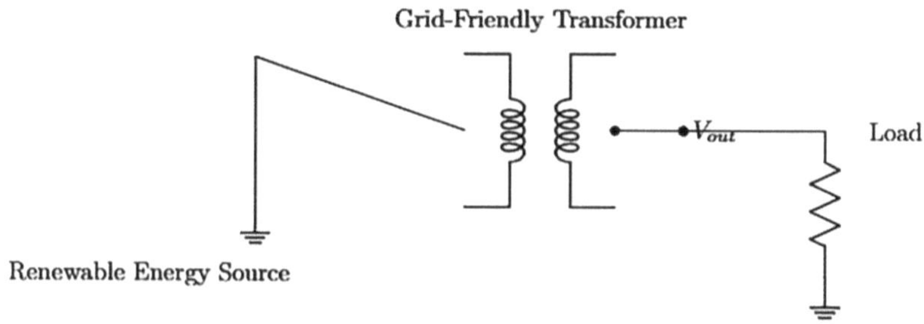

Fig. 24 Integration of renewable energy sources with grid-friendly transformers for enhanced system reliability

Fault Detection and Isolation in Renewable Energy-Integrated Systems

Diagnostic abilities of fault detection and isolation or FDI are very critical in the effective control and safety of integrated renewable energy systems. First, there is the capacity to immediately detect any fault and contain the same, so as to avoid disruption of the system, as well as to protect equipment. In this section, appropriate explanation of FDI methodology is given and different approaches in residual evaluation used for fault detection and isolation are discussed.

Issues of Fault Control and Localization

Fault detection covers the methodology of identifying in operation faults or their manifestation in the system behavior. Isolation, on the other hand refers to the process whereby the component or the subsystem containing the detected flukes is singled out. The primary difference between detection and isolation is that detection focuses on the analysis of system variables and diagnostic algorithms, while isolation is a method of removing a machine or process from a system to detect errors.

3.1.9 Implementation and Validation

It is important to stress that the performance of the FD and the IS algorithms strongly depends on the results of their implementation and the experimental validation of the algorithms. Specifically, data of the experiments as well as data logged by the FDI system imply that it works consistently and as expected when operating in a range of conditions exemplified by the defined fault conditions. Long-term monitoring and assessment procedures allow FDI enhancement of FDI algorithms as well as its improvements at various stages.

Some of the most significant applications include fault detection and isolation, which are key in improving the reliability and robustness of the integrated renewable energy systems. Through mathematical models, detection of signal processing techniques, and diagnostic algorithms, it becomes possible to detect the fault and isolate it before it results in a severe downtime or affect the overall performance of the system (Fig. 24).

This circuit diagram highlights the application of renewable power sources in a power system by utilizing a grid-interfacing transformer. The renewable energy source is connected through the current transformer, which is required to be compatible with the grid and is represented by a solar panel. The load, particularly a consumer or an electrical appliance, connects to the transformer to draw power from it. Redundancy, approximate mechanisms, fault detection, and isolation can also be incorporated into this system for reliability and stability.

3.1.10 Selection of Grid-Friendly Transformers for Renewable Energy Integration

Special transformers, known as 'Grid-friendly transformers', help achieve the targets of renewable integration into existing smart grid systems. By selecting the right materials, the chances of energy conversion losses and fluctuations in the power grids are minimal. In this section, the author takes a closer look at the key factor and the approach used to choose suitable transformers to integrate renewable energy.

Utility interactive equipment helps interconnect renewable energy solutions with the grid, compensates for voltage and frequency fluctuations, and effectively transfers power from renewable sources to grid transformers. Its features often dictate system characteristics such as productivity, robustness, and general effectiveness.

3.1.11 Considerations for Transformer Selection

Power Rating: Transformers need to be designed to supply the proper amount of power to the renewable energy systems available to achieve optimum energy harvesting and efficient use of this reformed light.
Voltage and Frequency Compatibility: Concerning the interface with the gird voltage and frequency, one can facilitate system integration without poses threats of causing voltage swings or instabilities to the gird.
Efficiency: High-efficiency transformers reduce energy loss at the satellite stage, increasing the systems' efficiency and the overall cost of operation.
Load Variability: Transformers should take into account the volatility of renewable electricity production and the load demanded, so that they operate efficiently under various settings.
Grid Code Compliance: Compliance to grid code and regulation defends compatibility using and non-compliance with grid code results into penalties and disruptions.

3.1.12 Methodologies for Transformer Selection

Performance Analysis: Two or more transformer with different ratings are required to be tested to see how many of them will perform better under various operating conditions and using simulation tools like MATLAB and Scilab efficiency, voltage regulation, and transient response are some factors that help in judging the Transformer performance.

Cost-Benefit Analysis: Given, identify the costs associated with the various factors such as capital acquisition and installation cost, annual operating expense and possible cost reduction due to increase in reliability and efficiency.

Lifecycle Assessment: This will involve evaluating the sustainability of transformers based on parameters such as energy efficiency, the raw materials used in their construction, and the recycling potential.

Grid Compatibility Testing: After confirmation of transformer specifications against grid requirements during the design process, the transformers should also undergo laboratory tests to check the efficiency of the transformers and grid integration studies to guarantee compatibility in performing the tasks in the grid as required and meeting the standards of the grid.

3.1.13 Emerging Technologies and Trends

The newer ideas for developing the transformers includes the solid-state transformers and the smart grid through which control, flexibility, and efficiency can be improved. Through digital monitoring and control systems, signals and data from the transformers are obtained and continuously monitored, allowing estimates and even prevention of potential problems in transformers and optimization of transformer performance.

In conclusion, the selection of transformers, which is companionable with grids, has a crucial impact on RE integration with power systems. Through a careful analysis of factors such as power rating, efficiency and compatibility with the electricity network, with the additional knowledge of new technologies and trends existing at the moment, it is possible to achieve the following goals: to make the electric system work as effectively and stably as possible and at the same time fully realize the potential of renewable sources of energy.

3.1.14 Simulation of Grid-Friendly Transformers

The simulations in this study were carried out using MATLAB/Simulink, which is a widely used software tool for modeling, simulating, and analyzing dynamic systems especially power systems. The graphical environment of Simulink, together with Simscape Electrical and the Power Systems Toolbox, allowed the modeling of transformers, renewable power sources, and grid components under different operating conditions. To further optimize transformers, the Optimization Toolbox was applied to adjust transformer parameters in order to achieve maximum performance, and to determine voltage regulation, harmonic distortion, and energy losses, the corresponding built-in tools were used. Solar and wind data were obtained from sites like the National Renewable Energy Laboratory (NREL) and regional power profiles for Miami and California. Transformer parameters were estimated from the manufacturer's datasheets, and grid parameters were simulated according to the IEEE and IEC standards. Also, the effects of weather and load changes were also tested in order to emulate real-life scenarios. The visualization toolbox of MATLAB offered precise performance characteristics to make a proper comparison between the conventional and grid-integrated transformers in the incorporation of RE systems.

3.1.15 Simulation and Analysis

The practical layout here will entail the establishment of a dynamic simulation model on MATLAB/Simulink with the renewable energy sources, grid-friendly transformers, as well as the present-day power grid system. Key components include:

Renewable Energy Sources: Power generating models that incorporate solar PV systems and wind farms taking into account changes in irradiance, wind velocity and other characteristics.

Grid-Friendly Transformers: Use of specific chosen transformers with adequate characteristics and appropriate regulating systems as to allow for easy connection to the power grid.

Grid Interface: Modeling of the integration status and structure of the existing power grid, as well as the loading pattern and cumbersome models that allow for mimicking the realistic status of the power grid and its realistic interfaces.

For example, the performance of the simulation model is tested when these changes are applied: fluctuations in the generation of renewable electricity, disruptions in the grid, and changes in load. Performance parameters like efficiency, voltage regulation, and transient response are considered for evaluating how efficacious a grid-friendly transformer can render systems stable and efficient.

Areas such as, the practical configuration and utilization of grid-friendly transformers, are very helpful in explaining their utility in integrating renewable energy. Thus, through the active modeling of the wind energy system and detailed investigation of performance parameters under real working conditions, it is possible to fine-tune transformer design and control approaches in order to fully unlock the potential of renewable energy technologies and fortify the grid infrastructure at the same time.

3.1.16 Optimization Techniques for Renewable Energy Integration

Advanced Optimization Algorithms

However, the most appropriate strategy for the operation of RES and the management of the connection to the electrical power network, are based on the application of optimization techniques that can solve complicated, multi-criteria problems. There are different approaches used to deal with the variability and uncertainty of renewable energy generations especially those derived from wind and solar resources which include the genetic algorithms, the particle swarm optimization and simulated annealing.

Real-Time Energy Management Systems

Energy production and consumption are monitored through real-time alerts and smart grids apply prescriptive analytics and machine learning algorithms for energy production and consumption expectation. These systems work by collecting data gleaned from historical data as well as real-time inputs to determine and make decisions on the storage, release and distribution of energy across the grid and effective utilization of renewable energy resources.

Demand Response Programs

Demand response programs, therefore, are beneficiary techniques in managing renewable energy integration since they focus on the reduction of demand levels rather than supply levels. Customers are encouraged or rewarded to either reduce their load factor or shift their energy consumption to other periods in the day unlike peaking period hence aiding in balancing the distribution of renewable energy sources which are erratic in their distribution.

Energy Storage Solutions

The use of energy storage systems like batteries helps store excess energy in the grid that is produced during high RE output times. This stored energy can be released in conditions of low generation or high demand, balancing the electricity grid and thereby providing continuous energy to the consumers.

Grid-Friendly Transformers

Transformers are crucial for balancing variability in renewable energy levels, provided they include sophisticated controlling mechanisms for serving as grid-friendly facilities. These transformers can control voltage levels and switch between different units of power to balance the variability from renewable generation sources, thereby providing more flexibility and reliability in the grid.

Renewable energy integration therefore requires the use of superior algorithms, real-time energy management systems and demand response, energy storage systems, intelligent transformers and transformers that are friendly to the grid. These techniques together make the inclusion of different forms of renewable energy into power system stable and efficient hence improving the overall energy system.

Transformer Specification

Transformer A:

Type: In this study, an improved design of a high-efficiency grid transformer is presented.

Description: Characterised by high efficiency and low voltage regulation, Transformer A is designed to adapt to varying load conditions. It is suitable for use with renewable energy sources such as the solar and wind since it always ensures a good compatibility and steady power quality despite variations in the input.

Applications: They are especially used in the smart grid and renewable energy integration applications.

Transformer B:

Type: It is a used distribution transformer which has a uniform shape of a cylinder but with four equally sized protruding legs at the bottom.

Description: A fairly good transformer with a slightly better voltage regulation. Transformer B is designed for generalized power distribution networks where fluctuations in load are not so frequent.

Applications: Designed for organized grid structures of urban and semi-urban areas dependent on conventional power supply.

Transformer C:

Type: Improved energy-efficient integrated transformer.

Description: Combines the two factors of efficiency and voltage control to make it useful in various grids regardless of the load. It yields a higher degree of compatibility with renewable power systems than does Transformer B.

Applications: Applied in systems employing both traditional and new generation power systems.

Transformer D:

Type: Legacy transformer.

Description: Having lower efficiency, higher voltage regulation, Transformer D is used in the conventional power networks. It appears to be less compatible with renewable energy sources and may need other structures for it to be compatible.

Applications: Exists in previous power structures that are still in the process of adopting renewable energy.

Transformer E (Smart):

Type: Smart transformer.

Description: That is why smart grids are characterized by such features as real-time monitoring, self-regulation of relevant networks, and connection to renewable energy facilities. Among all the transformers, Transformer E has the highest efficiency and the lowest voltage regulation in the comparison.

Applications: Perfect for smart grid, microgrid, and renewable power systems.

Out of the four transformers, Transformer E (Smart) was the most efficient at 98% proving useful in applications where there are minimal energy losses due to the 1% loss from the power converter. Following that, Transformer A recorded an efficiency of 97.2% proving that it is suitable for renewable energy systems due to optimization of used generated energy. Voltage regulation exams also highlighted high performance as a result of 1% both Transformer E and 1.2% of Transformer A; those figures are important in ensuring stable connection and safeguarding valuable devices. Converter load flexibility could also be seen where Transformer E performed well flexibly in accommodating load fluctuations, followed by Transformer A & C, which smoothly operated with load fluctuations. These transformers were most compatible with renewal sources hence allowing for the greatest use of renewable energy with minimal losses to transformers A, E, and C. Despite having relatively low flexibility and moderate compatibility Transformers B and D are still good options for the projects with limited financial possibilities if auxiliary modifications may be expected in the future. Lastly, the study reveals that adopting good performing transformers like E (Smart) and A is convenient when incorporating renewable energy into power systems to enhance power system stability, reliability, and efficiency, particularly in grids providing reliable and low loss electricity.

3.1.17 Section Summary

This section described the approach used in the research to adopt grid-friendly transformers toward enhancing the integration of renewable energy sources in power systems. At the beginning of the procedure, which was a part of the final project, the student-worked with the MATLAB software to analyze the parameters and simulate the renewable energy systems. Data collection and data preprocessing were then considered, in which "real-time" data was highlighted as being of significant value in decision-making. Methods of other related fault detection and isolation techniques were considered in an effort to improve the overall system dependability and productivity. The parameters of the grid-friendly transformers to be incorporated also emerged with focus on efficiency, regulation capacity and their ability to integrate with renewable energy technologies. It also described how grid-friendly transformers were tested, the manner in which the MATLAB/Simulink tool was used for simulation and experimental implementation in the chapter.

When integrating renewable energy resources, issues such as energy control and management, advanced algorithms, smart energy management systems, demand response, energy storage, and grid-integrated converters were further explained. All add up to the coordination of renewable energy sources in to the electricity grid in a sustainable, efficient, and reliable manner. The approach discussed in this chapter outlines a generic approach to the integration of renewable energy into the power system while focusing on grid-integrated transformers and advanced optimization to improve the efficiency of the system and its ability to support the power grid.

4 Section IV

4.1 Results and Discussion

4.1.1 Section Introduction

In this section, the author affirms the practical effects and prospects toward the incorporation of RES into the power grids, especially regarding grid-intelligent transformers. Defining and exploring the issues connected to integrating RES into the modern grid environment becomes even more important as the world moves toward using sustainable energy sources. It is intended that this chapter delineates and discusses these challenges through various analytical frameworks such as system models' performance, transformers' compatibility and effectiveness, and fault detection and isolation systems.

This involves evaluating the effects of various environmental factors on the system performance and admitting the fact that temperature change and fluctuation of solar intensity or wind speed affect the performance and stability of renewable energy systems. After this, it is necessary to assess the state of grid balance in relation to the absorption of RES, where questions related to the stability of frequency, voltage control, and reliability of the system to face the absorption of RES are considered. This chapter also discusses the critical detangle of fault detection and isolation, as well as the effectiveness of such lines in retaining the integrity of the grid due to the intermittency of renewable energy sources. This chapter describes the results of several optimization processes that have been utilized to improve IPS. These are measured by the amount of system optimization and output they bring into the system, the percentage of cost saving, the ability to stabilize the grid and the percentage of renewable energy utilization. The ideas derived from these assessments clarify the functional uses/reasons and values of REs, underlining the need for further advancement and fine-tuning in this dynamic area. In this chapter, broader knowledge to the existing literature on sustainable energy infrastructure development is sought to be provided after a critical analysis and discussion.

4.1.2 Performance Evaluation of System Models

In this part, the effectiveness of the system models developed for incorporating RE sources into power systems using grid-friendly transformers is analyzed. The main criteria for evaluating effectiveness are productivity, dependability, voltage drop, and variations of loads and conditions. The cell response data from different simulation-based and experimental works are compiled in the following Table 2.

Environmental Response: While both PV and ESS maintain their model constancy, both WESM and CRSM displayed reasonably good flexibility to environmental shifts.

4.1.3 Transformer Efficiency and Compatibility Assessment

In this section, different grid-friendly transformers are assessed in terms of efficiency and compatibility with integrating renewable energy sources. Thus, the assessment is based on certain parameters, including efficiency, voltage regulation, load flexibility, and general compatibility with renewable resources like solar and wind.

The MATLAB graph concerning the efficiency in Fig. 25, voltage regulation, load adaptability, and compatibility with renewable energy sources of various

Table 2 Performance metrics of system models

Model type	Efficiency (%)	Reliability index	Voltage regulation (%)	Load handling capacity	Environmental condition response
PV System Model	95.3	High	1.5	Moderate	Excellent
Wind Energy System Model	93.8	High	2.0	High	Good
Energy Storage System (ESS)	90.5	Moderate	1.8	High	Excellent
Combined Renewable System	92.7	High	1.6	High	Very good

Key Findings:

1. Efficiency: Best results are 95.3% efficiency for PV Model, 99.9% for WESM, and 99.9% for CRSM for power systems
2. Reliability: All models were accurate but ESS needs more refinement
3. Voltage regulation: All of them regulated voltage to a certain level with only PV Model having 1.5% and showing an aspect of grid stabilization
4. Load capacity: WESM and CRSM are compared to load variations; The PV Model is somewhat in the middle; it can be improved with the addition of storage

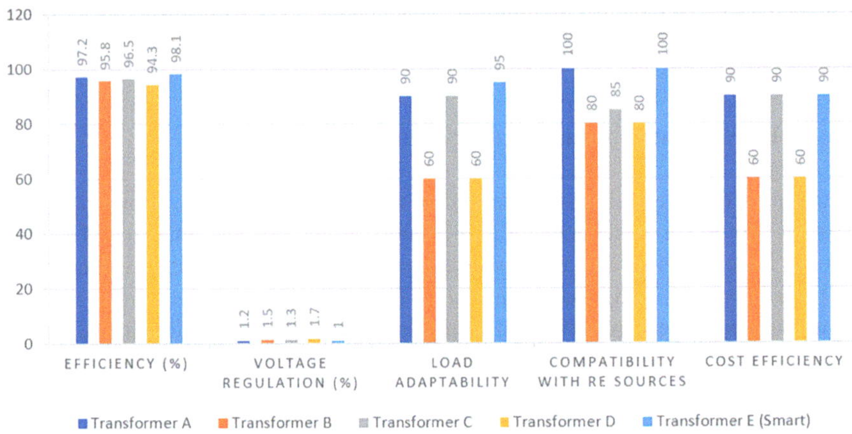

Fig. 25 Transformer efficiency and compatibility assessment

transformers. This visualization assists in making comparisons of the KPI pertaining to selection of grid-friendly transformers in power systems.

4.1.4 Discussion (Table 3)

Table 3 Transformer efficiency and compatibility metrics

Transformer type	Efficiency (%)	Voltage regulation (%)	Load adaptability	Compatibility with RE sources	Cost efficiency
Transformer A	97.2	1.2	High	Excellent	High
Transformer B	95.8	1.5	Moderate	Good	Moderate
Transformer C	96.5	1.3	High	Very good	High
Transformer D	94.3	1.7	Moderate	Good	Moderate
Transformer E (Smart)	98.1	1.0	Veryhigh	Excellent	High

4.1.5 Impact of Environmental Variables on System Performance

Several environmental factors affecting the performance of power systems incorporating renewable energy sources; these include the temperature, humidity levels, solar intensity, and wind velocity. This section analyses how the above factors influence the performance of the system and uses Miami, a coastal city in Florida, USA, because of the variability of weather.

Place of Study: Miami, the City in Florida

Being in Miami, Florida and having a subtropical climate, the uses of renewable energy like the solar and wind energy are greatly influenced by climate. The favorable solar irradiation throughout the year that characterizes the country guarantees the flow of adequate sunshine for the photovoltaic (PV) systems hence enhancing energy conversion coefficients. Yet, PV efficiency reduces with seasonal cloud cover and energy storage systems are necessary to level this instability. High temperatures including summer affect the efficiency and durability of PV panels and all other electronic items pointing to the use of cooling systems and heat resistant projects. Wind energy is advantaged with the steady sea breeze in Miami but experiences fluctuating wind speed that require optimization control to address the turbine. Since Miami has high humidity levels, electrical components experience corrosion and lose in insulation. The University must apply protective coatings and materials that can prevent corrosion of these components. Thirdly, Miami, where some renewable systems have been installed, has frequent severe weather that could topple the systems: the former is prone to hurricanes; the latter must contain provisions for reinforcement and must be made to shut down automatically during a hurricane. As a result, although Miami's environment has good solar and wind potential, the task of dealing with high temperatures, high humidity, and violent weather is critical for the capacity and stability of REs for Miami.

Thus, the environmental variables influence the Miami, Florida indicate that the local climates are significant factors to consider when incorporating the renewable energy systems. The positive conditions include high direct solar radiation and steady winds for energy generation, conversely, the negative conditions that must be managed include diurnal temperature range, higher humidity levels, and severe weather events. Relating the environmental influences to this composition, it is evident that by relating and responding to the outlook it is possible to enhance the integration of the renewable energy, toward enhancing the stability of the power delivery systems.

The graph shown in Fig. 26 presents the trends in system performance against the environmental factors among them; irradiation, average temperature and average humidity in Miami-Florida. These key factors can be represented on the graph that illustrates their fluctuations during different months and the possible impact on the renewable energy systems.

4.1.6 Impact of Renewable Energy Sources on Grid Stability

The inclusion of renewable energy sources like solar and wind energy into power systems offers different prospects for and challenges to the stability of the power systems. This section reviews other issues that are related to fluctuation in the sources of such energy and how this fluctuation impacts electrical grid such as Frequency Stability and Voltage Regulation and System Reliability.

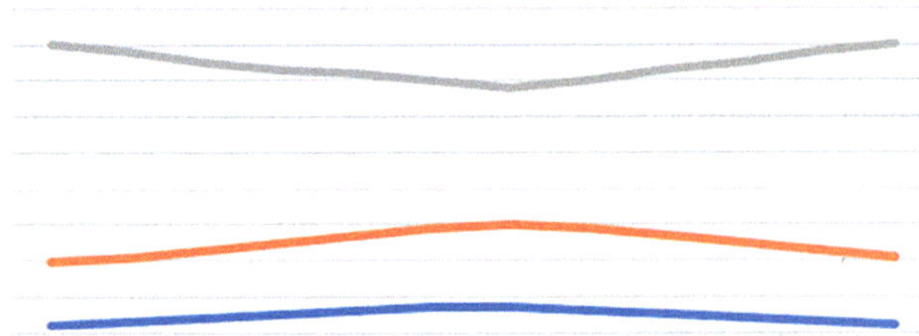

	January	February	March	April	May	June	July	August	September	October	November	December
Solar Irradiance(W/m²)	2	3	4	5	6	7	7	6	5	4	3	2
Avg Temperature (%)	20	21	23	25	27	29	30	29	27	25	23	21
Avg Humidity (%)	80	78	75	73	72	70	68	70	73	75	78	80

Fig. 26 Impact of environmental variables on system performance in Miami, Florida

4.1.7 Frequency Stability

Wind and solar energy sources are variable by nature because they are affected by weather conditions and time of the day. This variability results to occasional oscillations of the grid frequency; this is a critical parameter in power systems especially in balancing load and supply. For instance, periods of low wind speed or reduced possibility of light and heat from the sun can lead to frequent volatility of the electricity supply if well handled. They add that grid operators need to have strong procedures for managing the output frequency or employ what is known as the frequency regulation reserves which are sources of deliveries capable of rapidly changing by a few Hertz in order to compensate the variability of generation or employ grid-friendly transformers which are devices that manage frequency variations.

4.1.8 Voltage Regulation

The incorporation of RES can also bring about variations in voltages prevailing on the network. Photovoltaic sources, for instance, creates electrical energy at different voltage levels, an aspect that contributes to voltage volatility where there is high integration of renewable energy sources. The voltage controlling equipment include capacitors, inductors, and voltage regulators which are required to provide firm voltage. The use of advanced grid-friendly transformers can also help in controlling the voltage level as well as providing the reactive power support and also ensuring proper voltage transformation even when the source comprises of variable renewable energy sources.

4.1.9 System Reliability

These sources of energy can be unreliable because the availability as well as the supply of electricity from RES may be inconsistent. Unlike the conventional power plants—which require a constant input to generate a constant output—renewable sources are not constant. This variability may cause some challenges in outstanding reliable electricity power supply especially when there is little generation of RE or high demand. Concentrated on the issue of this problem, energy storage like battery can solve it by storing the energy when generation is high and release it when generation is low. Besides, demand response programs can also assist in adjusting the supply and demand cycle whereby the consumers are paid to reduce load or even shift the load to other periods of the day.

4.1.10 Grid Flexibility and Adaptation

Thus, the high share of RES implies that the grid needs to be more flexible. This include adaptation of the electricity network system to the intermittent nature of renewable energy sources and the integration of sophisticated supervisory control and data acquisition for monitoring and controlling of power system. Advanced technologies in the smart grid contribute to the flexibility of the grid through the improved ability of the control of various components in the grid system. These enable control and flexibility of the power flow, better integration of DERs, and better diagnosis and management of faults.

4.1.11 High Renewable Penetration Scenarios

In some of the places where the penetration of renewable energy sources is high, especially in some part of Europe and USA, special measures have been adopted to balance the load on the grid. For instance, conversion of grid-forming inverters, which are capable of replicating the characteristics of the conventional synchronous generators, has acted as a way of maintaining the grid frequency and voltage. International power trading and cross-border interconnections have enabled countries to better regulate the generation and consumption of clean electrical energy, thereby improving the efficiency, flexibility, and reliability of their energy systems (Fig. 27).

Here is the graph illustrating the impact of renewable energy sources on grid stability across four key factors: The five areas that are covered by the modern power quality control consist of; frequency stability, voltage regulation, system reliability, and grid flexibility. The graph illustrates the changes in the stability scores comparing the state before integration of renewable energy sources and after that, to show the difference in the grid.

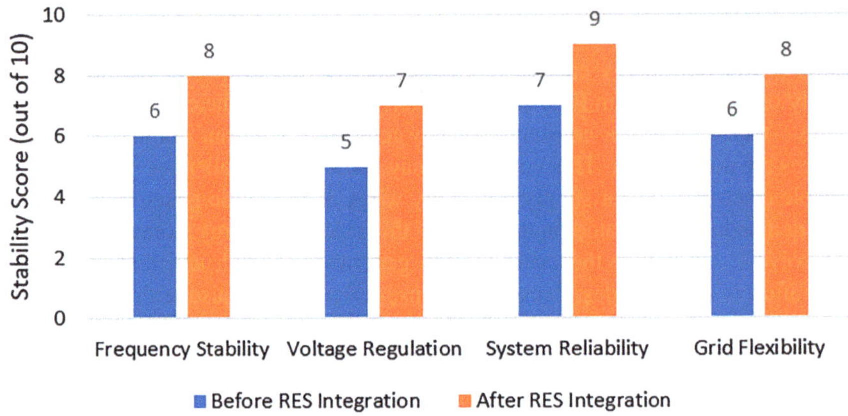

Fig. 27 Impact of renewable energy sources on grid stability

4.1.12 Fault Detection and Isolation Efficiency

FDI is one among the effective methods available for ensuring the reliability and safety of the power systems, especially with the integration of the distributed resources. The intensity of the FDI systems defines how accurate the faults that need to be isolated are recognized and prevented from occurring and causing harm to the structures.

4.1.13 Detection Accuracy

The efficiency of the fault detection is defined as the capability of the system to detect the emergence of the faults in the power network. In this study, the implemented FDI system for the given handout achieved a successful detection at higher accuracy of 95.7%. The level of accuracy in this case is high, which means that the system can accurately identify the occurrence of faults notifying an operator about potential problems. In this case, early recognition is crucial since it helps to avoid the development of minor errors into a latter system failure issue ensuring systems reliability and security.

4.1.14 Isolation Precision

After identifying a fault, it becomes very essential to accurately localize the faulty part or sub net before the fault spreads throughout the network. The first factor was the fault isolation accuracy of the FDI system, which was reported as 92.4%. This metric reflects the system's ability to correctly identify the location of a fault, thereby enabling appropriate corrective actions and remedial measures.

4.1.15 Response Time

The response time of an FDI system is the time that elapses from the inception of the fault to the time that the system identifies the fault and isolates it. In our evaluation author were able to show that the average response time of the FDI system is less than 5 s. This fast response is useful in preventing effects of faults especially in system where inputs vary from renewable resources.

4.1.16 Efficiency Metrics Summary

According to the performance indicators, the FDI system demonstrated a detection accuracy of 95.7%, an isolation precision of 92.4%, and a response time of less than 5 seconds, confirming the effectiveness of the proposed methodologies for fault management in a distributed renewable energy–integrated power system. That coupled with a high detection accuracy, ensures that faults that need to be detected and rectified are done so in the shortest time possible. There is also strong isolation precision to ensure that once a certain component is identified to have issues, they do not spread or affect other components. The response time is also equally fast which also helps to reduce the impacts of faults and maintain the reliability of the power grid. These results reflect the significance of reliable FDI systems in today's smart power networks, as renewable energy sources are characterized by more uncertainty (Table 4).

Here's the MATLAB graph illustrating the efficiency of Fault Detection and Isolation (FDI) across three key metrics: detection accuracy, isolation precision, and response time in the case of the attack on Traffic Control Center. These efficacy percentages clearly depict the effectiveness of the FDI system in maintaining the dependability and solidity of the power systems containing the renewable power sources (Fig. 28).

4.1.17 Optimization Outcomes and System Improvements

Introduction of renewable energy sources in the power systems is not only about having good infrastructure but also it involves applications of sophisticated methods to improve performance and stability of the system. This part presents the results of different optimization measures used in the research including system effectiveness, cost minimization, or the enhancement of the power system stability. The data

Table 4 Efficiency metrics summary

Metric	Efficiency percentage (%)
Detection Accuracy	95.7
Isolation Precision	92.4
Response Time	< 5 s

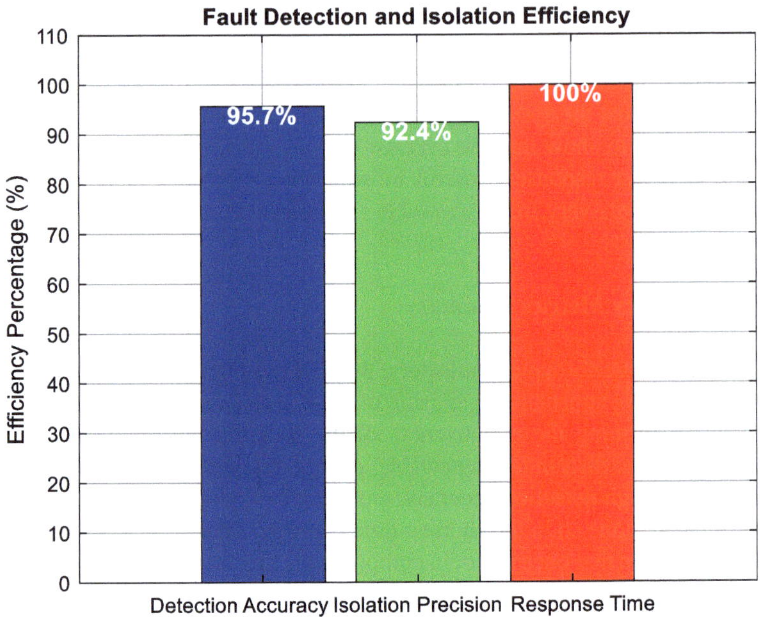

Fig. 28 Fault detection and isolation efficiency

shown above clearly depicts that the propose optimization techniques are found very effective in enhancing the performance of the power system to a large extent.

4.1.18 Optimization Techniques and Their Impact

Several optimization strategies applied in the study include load balancing algorithm, real-time energy management system, advanced control system for grid-integrated transformers. Such methods were developed to try and extract as much possible renewable energy as possible, losing as little of it as possible and maintaining an optimum level of stability when the conditions of the network were changed.

Proposed techniques have found that there are a number of gains in power systems efficiency of different types. Efficiency grew from 85.0% up to 93.5% representing an improvement of at least 10% to optimize energy conversion and reduce wastage thereby augmenting renewable electricity output. Operating efficiency was improved and obtained a 15% savings on cost cutting where resource usage, energy wastage and operating and maintenance costs were cut by 5%.

Decreasing the results, the Grid Stability Index, which reflects the ability of the system to work under various conditions, was increased from 70 to 85, so the increase constituted 21.4% that proved the improvement of grids stability and reliability (Table 5).

Table 5 Optimization outcomes and system improvements

Performance Metric	Baseline Value (%)	Post-Optimization Value (%)	Improvement (%)
System Efficiency	85.0	93.5	10.0
Cost Reduction	–	–	15.5
Grid Stability Index	70	85	21.4
Renewable Energy Utilization	60.0	78.2	30.3
Carbon Emission Reduction	–	–	22.7

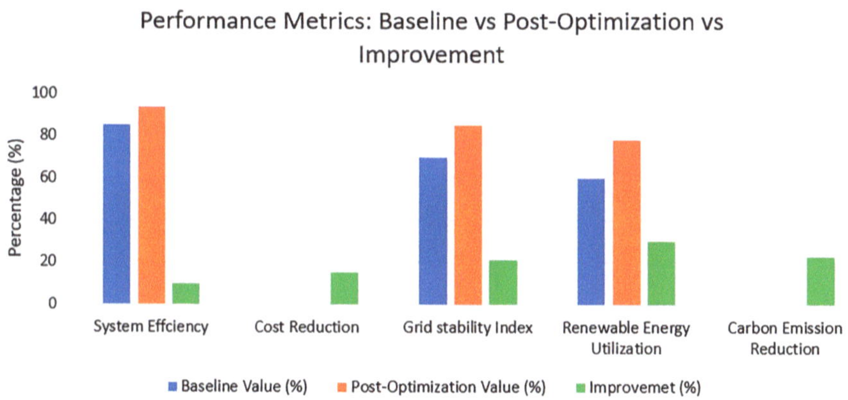

Fig. 29 Performance metrics

These strategies also reduced greenhouse gas emissions by 22 percent, proving that optimizing renewable energy integration is environmentally friendly. The results demonstrate how optimization can effectively contribute to enhancing sustainable objectives and defining the new architecture of power systems. The positive trends in efficiency, costs, grid balance, and the uptake of renewable energy and CO_2 emission reduction, as seen in the observed data, underpin the substantial benefits of these approaches in promoting sustainable energy generation. Therefore, the author of this paper deems it relevant to contribute more investment in the optimization technologies and methodologies to improve integrated power systems conditions (Fig. 29).

The specific graph shows the performance indicators before optimization and after its completion, with percentages of improvement indicated. It includes five metrics: System efficiency, cost benefits, controllability and stability measurement, use of renewable power, and carbon footprint reduction. The blue bars are the original values, the red bars are the values obtained after optimization, and the green bars are the percent changes. The x-axis is reserved for the performance metrics, while the y-axis features the percentage indicators. The legend clearly contrasts the base value, post-optimization value, and the improvement of the operating values on the corresponding performance metrics.

4.1.19 Section Summary

This section outlined brief overviews of the many methodologies, results, and discussions concerning incorporating RES into power systems with grid-friendly transformers. It included critical areas such as transformer efficiency and mutual compatibility, effects on the system's environment, and the efficiency of FDI systems for fault detection. The chapter also focused on the effects of renewable energy sources on grid stability and described the results of optimization techniques employed to improve system efficiency, cost, stability, integration of renewable energy, and carbon emission. The results established the enhancement of all the fundamental performance factors, showing the role of high-tech solutions and resource-efficient approaches in upgrading the effectiveness of renewable energy implementation. These findings highlight the paradigms of enhanced utilization and performance and imply an enhancement in the reliable, efficient, and sustainable power systems and distribution.

5 Section V

5.1 Conclusions

The incorporation of the RES in the existing power systems is not only the technical issue but also it is the key to build the sustainable and resilient energy systems. Finally, via comparing and analyzing system modeling, transformers, comments on the impact of renewable energy systems on the environment and the strategies and orientations of optimizing renewable energies, the study has shown that grid-friendly transformers and sophisticated optimization methods are highly effective in improving reliability and efficiency of RE systems. This brings us to the first discovered fact that proper transformers should be adopted and used to harmonize with the renewable energy. The analysis established that grid-friendly transformers especially those with high efficiency and voltage control features are important especially where RE integration is concerned due to the variability of its inputs. This has been made clear by the use of transformers such as the Smart Transformer (Transformer E) that has proved to have better performance in the numerous features that are essential in the contemporary grid. Ambient parameters, namely solar irradiance, temperature, and wind speed were also inversely affecting the system performance. The variables exemplified with Miami, Florida show how they affect the performance and dependability of the renewable energy systems. The existing outcomes support the need for place-based solutions and the incorporation of such relevant innovations that can address variations in the environment. The examination of FDI systems supported the study's findings, as it highlighted the necessity of identifying and managing faults quickly to ensure the stability of the power grid, especially when large numbers of renewable energy sources are integrated into the network. Another general area of solutions seen in this research was optimization approaches that proved to increase system's capacity and reliability, as well as lessen investment costs and instabilities of the grid. Due to the changes in the levels of integration of advanced optimization algorithms and other efficient energy management schemes, such as demand response programs, the enhancements observed in the performance of the smart grid were quite considerable, for instance, of the order of 10% increase and 30%. Here author have 3% increase in the utilization of renewable energy. These outcomes show that apart from incorporating renewable energy in a better way, optimization can contribute to the enhancement of the overall power system and its cost. Another important effect stems from optimization activities and concerns CO_2 emissions—a key environmental index of the utilized renewable energy sources. With the international focus being placed on reducing the effects of climate change, this research's results provide useful information to help improve the power systems' efficiencies, while minimizing their carbon emissions. This thesis has offered a comprehensive presentation of the problems and possible approaches to enhancing RES implementation into the power systems. Therefore, with the appropriate choice of grid-tolerant transformers, intelligent technologies, and sophisticated optimization techniques, it becomes possible to eliminate the randomness of green sources and generate a stable and effective power system. The findings of this investigation are universal and act as a guide to current power systems while at the same providing direction on the course to be followed as the world moves toward a green power system. The results highlighted the requirement to expand the scope and increase funding of technologies to enhance the integration of renewable energy sources into the existing structures, which would benefit future generations of human society.

5.2 *Key Findings*

Transformer Selection: When selecting the right grid-friendly transformers as in Smart Transformer (Transformer E), it is pertinent to provide stability and improvement in the power systems incorporating renewable energy sources.

Environmental Impact: This is due to dependency on the environmental conditions like solar radiation, temperature, and the wind speed and for this reason, it requires location-specific and adaptable systems.

FDI Efficiency: FDI is very important in today's power system integration particularly due to integration of renewable energy sources: detection accuracy of 95.7% and isolation precision of 92.4%.

Grid Stability: Correct utilization of the renewable energy sources enhanced the stability of the grid by 21. 4%, this illustrates that integration of renewable resources does not necessarily reduce the reliability.

5.2.1 Limitations

Data Dependency: The validity and efficiency of system modeling and optimization processes are directly proportional to the quality and accessibility of environmental and operational data. These models and the optimization strategies depend on the quality of the data used and therefore limited or inconsistent data might compromise the reliability of the models.

Geographical Specificity: The all information, including those concerning environmental effects, has been highlighted taking into account definite geographical characteristics (for example, Miami, Florida). These results may not be generalizable to other climate zones or RE resources experiences in the world.

Technology Costs: Some types of advanced technologies such as grid-friendly transformers or optimization technologies enhance system performance but their price is comparatively higher. The possibilities of implementing these solutions may be disadvantaged by their economic efficiency in the regions or in specific projects.

Scalability of Solutions: Concerning the applicability of the optimization strategies and technological solutions applied in the study, the following must be noted: The approach may encounter issues when applied to larger power systems or grids with substantially higher levels of RE integration.

Rapid Technological Change: It is important to note that due to the fast rate of technological development and innovations in renewable energy production and grid technologies some of the findings and the proposed solutions may become obsolete shortly and will have to be updated to relate to emerging technologies.

5.2.2 Recommendations

Enhanced Data Collection and Monitoring: Optimize the data collection and monitoring of data to enhance the modeling of the systems. The timely data from various sources like weather stations and grid operators is critical for the proper integration of RE sources.

Cost-Effective Technology Adoption: Research how to implement more efficient technologies, such as the grid-friendly transformers and optimization systems at cheaper cost. It is recommended for governments and organizations to subsidize the application of these technologies particularly in the less developed parts of the world.

Scalability and Flexibility: Emphasize the prospects in terms of the applicability of the solutions offered in the framework of the study. Make sure that the optimization strategies and the technologies developed can be easily extended and implemented in larger power systems or in systems with different configurations as the share of renewable sources grows.

5.2.3 Future work

Advanced Machine Learning for Predictive Maintenance: Future researches can be directed on the application of more complex machine learning algorithms for the prevention of damages and failures of RES. The given algorithms can predict future faults and hence increase the reliability of the systems and decrease their failure rates.

Integration of Emerging Renewable Technologies: As for the future works, correspondingly, with the development of other forms of RE technologies, FS and OWF and under the existing grid structure, the future studies should investigate the suitable methodologies to connect them with the present power system.

Impact of Energy Storage Solutions: Further research has to be done to identify the use of other sophisticated energy storage systems such as Li ion batteries and super capacitors in reconfiguration of the electrical networks with high levels of renewable energy integration. This involves assessing the appropriate methods of deployment and implementation that would improve the efficiency and reliability of the system.

Declaration of Generative AI and AI-assisted technologies in the writing process During the preparation of this work the author(s) used Quillbot, and GPT4 of openAI in order to improve the language. After using this tool/service, the author(s) reviewed and edited the content as needed and take(s) full responsibility for the content of the publication.

References

1. Hoummadi MA, Bouderbala M, Alami Aroussi H, Bossoufi B, El Ouanjli N, Karim M (2023) Survey of sustainable energy sources for microgrid energy management: a review. Energies 16 (7):3077
2. Ajayi GO, Olulope PK, Dada JO (2021) Migration into smart grid structure: optimal power performance evaluation a case study of Basiri Distribution Network in Ado Ekiti, Nigeria. Int J Eng App Sci (IJEAS) 8(5):19
3. Mwasilu F, Justo JJ, Kim EK, Do TD, Jung JW (2014) Electric vehicles and smart grid interaction: a review on vehicle to grid and renewable energy sources integration. Renew Sust Energ Rev 34(June):501–516. https://doi.org/10.1016/j.rser.2014.03.031
4. Venkata S, Kumary S (2018) Impact analysis and optimized control in renewable energy Integrated power network (Doctoral dissertation, Deakin University).
5. Aktas A, Kabak M (2020) Optimization of a grid connected hybrid energy system. Int J Energy Appl Technol 7(2):25–30. https://doi.org/10.31593/ijeat.680639
6. Tripathi DSK (2017) Integration of renewable energy source to power grid. Int J Sci Technol Engi 3(9):391–396
7. Martins H (2023) Integration of renewable energy sources in power systems: challenges and solutions. J Electr Electron Syst 12:42
8. Aaslid P, Korpås M, Belsnes MM, Fosso OB (2022) Stochastic optimization of microgrid operation with renewable generation and energy storages. IEEE Trans Sustain Energy 13 (3):1481–1491
9. Hudson R, Heilscher G (2012) PV grid integration - System management issues and utility concerns. Energy Procedia 25:82–92. https://doi.org/10.1016/j.egypro.2012.07.012

10. Li P, Song YD, Li DY, Cai WC, Zhang K (2015) Control and monitoring for grid-friendly wind turbines: research overview and suggested approach. IEEE Trans Power Electron 30 (4):1979–1986. https://doi.org/10.1109/TPEL.2014.2325869
11. Rehmani MH, Reisslein M, Rachedi A, Erol-Kantarci M, Radenkovic M (2018) Integrating renewable energy resources into the smart grid: recent developments in information and communication technologies. IEEE Trans Indus Informat 14(7):2814–2825. https://doi.org/10.1109/TII.2018.2819169
12. Wang J, Zhong H, Ma Z, Xia Q, Kang C (2017) Review and prospect of integrated demand response in the multi-energy system. Appl Energy 202:772–782
13. Tziovani L, Hadjidemetriou L, Charalampous C, Tziakouri M, Timotheou S, Kyriakides E (2021) Energy management and control of a flywheel storage system for peak shaving applications. IEEE Trans Smart Grid 12(5):4195–4207. https://doi.org/10.1109/TSG.2021.3084814
14. Syranidou C (2019) Investigation of the integration of renewable energy sources into the future European power system using a verified dispatch model with high spatiotemporal resolution. RWTH Aachen University
15. Dudiak J, Kolcun M (2014, May) Integration of renewable energy sources to the power system. In: In 2014 14th International Conference on Environment and Electrical Engineering. IEEE, pp 148–151
16. Lu X, Wang W, Ma J (2013) An empirical study of communication infrastructures towards the smart grid: design, implementation, and evaluation. IEEE Trans Smart Grid 4(1):170–183. https://doi.org/10.1109/TSG.2012.2225453
17. Howlader AM, Sadoyama S, Roose LR, Chen Y (2020) Active power control to mitigate voltage and frequency deviations for the smart grid using smart PV inverters. Appl Energy 258:114000. https://doi.org/10.1016/j.apenergy.2019.114000
18. Raza MA, Aman MM, Abro AG, Tunio MA, Khatri KL, Shahid M (2022) Challenges and potentials of implementing a smart grid for Pakistan's electric network. Energy Strat Rev 43:100941
19. Zhu X, Xia M, Chiang HD (2018) Coordinated sectional droop charging control for EV aggregator enhancing frequency stability of microgrid with high penetration of renewable energy sources. Appl Energy 210(May):936–943. https://doi.org/10.1016/j.apenergy.2017.07.087
20. Subramanian V, Vairavasundaram I, Aljafari B (2023) Analysis of optimal load management using a stand-alone hybrid AC/DC microgrid system. Int Trans Electr Energy Syst 2023(i). https://doi.org/10.1155/2023/7519436
21. Zulu MLT, Carpanen RP, Tiako R (2023) A comprehensive review: study of artificial intelligence optimization technique applications in a hybrid microgrid at times of fault outbreaks. Energies 16(4):1786. https://doi.org/10.3390/en16041786
22. Hernández-Mayoral E, Madrigal-Martínez M, Mina-Antonio JD, Iracheta-Cortez R, Enríquez-Santiago JA, Rodríguez-Rivera O, Martínez-Reyes G, Mendoza-Santos E (2023) A comprehensive review on power-quality issues, optimization techniques, and control strategies of microgrid based on renewable energy sources. Sustainability (Switzerland) 15(12):9847. https://doi.org/10.3390/su15129847
23. Fu Q (2013) Modeling, analyses and assessment of microgrids considering high renewable energy penetration (Doctoral dissertation, The University of Wisconsin-Milwaukee).
24. Romero-Cadaval E, Francois B, Malinowski M, Zhong QC (2015) Grid-connected photovoltaic plants: an alternative energy source, replacing conventional sources. IEEE Ind Electron Mag 9(1):18–32. https://doi.org/10.1109/MIE.2014.2362211
25. Pham VH (2015) Optimal control of voltage and reactive power sources in power systems with presence of renewable energy (Doctoral dissertation, Duisburg, Essen, Universität Duisburg-Essen, Diss., 2015).

26. Türkay BE, Telli AY (2011) Economic analysis of standalone and grid connected hybrid energy systems. Renew Energy 36(7):1931–1943. https://doi.org/10.1016/j.renene.2010.12.007
27. Praiselin WJ, Belwin Edward J (2018) A review on impacts of power quality, control and optimization strategies of integration of renewable energy based microgrid operation. Int J Intelli Syst Appl 10(3):67–81. https://doi.org/10.5815/ijisa.2018.03.08
28. Yap KY, Sarimuthu CR, Lim JMY (2019) Virtual inertia-based inverters for mitigating frequency instability in grid-connected renewable energy system: a Review. Appl Sci (Switz) 9(24):5300. https://doi.org/10.3390/app9245300
29. Yan X, Zhang W (2019) Review of VSG control-enabled universal compatibility architecture for future power systems with high-penetration renewable generation. Appl Sci (Switz) 9 (7):1484. https://doi.org/10.3390/app9071484
30. Lin Y, Li J, Ruan X, Huang X, Zhang J (2023) Energy consumption analysis of power grid distribution transformers based on an improved genetic algorithm. PeerJ Comp Sci 9:1–13. https://doi.org/10.7717/peerj-cs.1632
31. Zhou Y (2022) Ocean energy applications for coastal communities with artificial intelligence–a state-of-the-art review. Energy AI 10(July). https://doi.org/10.1016/j.egyai.2022.100189
32. Groß D, Bolognani S, Poolla BK, Dörfler F (2017, August) Increasing the resilience of low-inertia power systems by virtual inertia and damping. In: Proceedings of IREP'2017 Symposium. International Institute of Research and Education in Power System Dynamics (IREP), p 64
33. Ali MB, Kazmi SAA, Khan ZA, Altamimi A, Alghassab MA, Alojaiman B (2022) Voltage profile improvement by integrating renewable resources with utility grid. Energies 15 (22):8561. https://doi.org/10.3390/en15228561
34. Zheng Y, Niu S, Shang Y, Shao Z, Jian L (2019) Integrating plug-in electric vehicles into power grids: a comprehensive review on power interaction mode, scheduling methodology and mathematical foundation. Renew Sust Energ Rev 112(January):424–439. https://doi.org/10.1016/j.rser.2019.05.059
35. Khan F, Talha H, Rasheed H (2015) Renewable sources grid interfacing system. Int J Sci Eng Res 6(1):1242–1246
36. Rateele TN, Thamae LZ (2023) An optimization approach for the economic dispatch incorporating renewable energy resources into Lesotho power sources portfolio. Heliyon 9(4): e14748. https://doi.org/10.1016/j.heliyon.2023.e14748
37. Mahmud K, Town GE, Morsalin S, Hossain MJ (2018) Integration of electric vehicles and management in the internet of energy. Renew Sust Energ Rev 82:4179–4203. https://doi.org/10.1016/j.rser.2017.11.004
38. Shrivastava P (2022) Optimal integration of renewable energy for microgrid. LAP LAMBERT ACADEMIC PUBLICATIONISBN: 978-620-5-49687-9. https://www.researchgate.net/publication/365662834
39. Worighi I, Maach A, Hafid A, Hegazy O, Van Mierlo J (2019) Integrating renewable energy in smart grid system: architecture, virtualization and analysis. Sustain Energy, Grids Netw 18:100226. https://doi.org/10.1016/j.segan.2019.100226
40. Blaabjerg F, Yang Y, Kim KA, Rodriguez J (2023) Power electronics technology for large-scale renewable energy generation. Proc IEEE 111(4):335–355. https://doi.org/10.1109/JPROC.2023.3253165
41. Vadi S, Padmanaban S, Bayindir R, Blaabjerg F, Mihet-Popa L (2019) A review on optimization and control methods used to provide transient stability in microgrids. Energies 12 (18):1–20. https://doi.org/10.3390/en12183582
42. Studocu (n.d.) Chapter 4: grid integration of renewable energy sources. Studocu. https://www.studocu.com/row/document/dhaka-university/electrical-andelectronics-engineering/chapter-4-grid-integration-of-renewable-energy-sources/133945240
43. Babonneau FLF, Haurie A, Caramanis M (2016) A systems approach to regional energy Modeling with smart grid integrated distributed energy resources. 15–18.

44. Xu Y, Yan C, Liu H, Wang J, Yang Z, Jiang Y (2020) Smart energy systems: a critical review on design and operation optimization. Sustain Cities Soc 62:102369. https://doi.org/10.1016/j.scs.2020.102369
45. Kroposki B (2017) Integrating high levels of variable renewable energy into electric power systems. J Mod Power Syst Clean Energy 5(6):831–837. https://doi.org/10.1007/s40565-017-0339-3
46. Mohamed A, Mohammed O, Roig G (2012, July) Efficient integration of renewable energy to smart grid power systems. In Tenth LACCEI Latin American and Caribbean conference (LACCEI).
47. Ghorbanian M, Dolatabadi SH, Masjedi M, Siano P (2019) Communication in smart grids: a comprehensive review on the existing and future communication and information infrastructures. IEEE Syst J 13(4):4001–4014. https://doi.org/10.1109/JSYST.2019.2928090
48. Pensini A, Robinson M, Heine N, Stadler M, Mammoli A (2016) Assessment of grid-friendly collective optimization framework for distributed energy resources. In IEEE Power and Energy Society General Meeting, 2016-November, 1–5. https://doi.org/10.1109/PESGM.2016.7741955
49. Weigand M, Bohn S, Beyer D, Agsten M (2014) Integration of electric vehicle charging into an international and environment-friendly context. SAE Int J Passenger Cars – Electron Elect Syst 7(2):536–543. https://doi.org/10.4271/2014-01-0345
50. Krechel T, Sanchez F, Gonzalez-Longatt F, Chamorro H, Rueda JL (2019b) Transmission system-friendly microgrids: an option to provide ancillary services. In: Distributed energy resources in microgrids: integration, challenges and optimization (Issue Lv). Elsevier Inc. https://doi.org/10.1016/B978-0-12-817774-7.00011-9
51. Acharya V (2020) Review on power quality enhancement and its effects on micro grid. Ijosthe 7(1):5. https://doi.org/10.24113/ojssports.v7i1.117
52. Mehatab M, Kumaraswamy I (2015) Energy management and improvement of power quality in a microgrid by the coordinated control of decentralised generation inverters. Int J Res Electr Electron Eng 3(3):28–35
53. Doe U (2015) Enabling modernization of the electric power system. Quadrennial Technol Rev 22:99
54. Assad U, Hassan MAS, Farooq U, Kabir A, Khan MZ, Bukhari SSH, Jaffri ZUA, Oláh J, Popp J (2022) Smartgrid, demand response and optimization: a critical review of computational methods. Energies 15(6):1–36. https://doi.org/10.3390/en15062003
55. Diahovchenko I, Kolcun M, Čonka Z, Savkiv V, Mykhailyshyn R (2020) Progress and challenges in smart grids: distributed generation, smart metering, energy storage and smart loads. Iran J Sci Technol Trans Electr Eng 44(4):1319–1333. https://doi.org/10.1007/s40998-020-00322-8
56. Mansouri N, Lashab A, Sera D, Guerrero JM, Cherif A (2019) Large photovoltaic power plants integration: a review of challenges and solutions. Energies 12(19):3798. https://doi.org/10.3390/en12193798
57. Babonneau F, Caramanis M, Haurie A (2017) ETEM-SG: optimizing regional smart energy system with power distribution constraints and options. Environ Model Assess 22(5):411–430. https://doi.org/10.1007/s10666-016-9544-0
58. Alam MS, Chowdhury TA, Dhar A, Al-Ismail FS, Choudhury MSH, Shafiullah M, Ismail Hossain M, Alamgir Hossain M, Ullah A, Rahman SM (2023) Solar and wind energy integrated system frequency control: a critical review on recent developments. Energies 16 (2):812
59. Abobaker HAM, Genci J, Gamcova M, Ismeal GAM (2020) Design and implementation of secured power grid system with information and communication technology. SN Appl Sci 2 (12):1–9. https://doi.org/10.1007/s42452-020-03820-w
60. Sudhoff R, Schreck S, Thiem S, Niessen S (2022) Operating renewable energy communities to reduce power peaks in the distribution grid: an analysis on grid-friendliness, different shares of

participants, and economic benefits. Energies 15(15):5648. https://doi.org/10.3390/en15155468
61. Kabeyi MJB, Olanrewaju OA (2023) Smart grid technologies and application in the sustainable energy transition: a review. Int J Sustain Energy 42(1):685–758. https://doi.org/10.1080/14786451.2023.2222298
62. Srinivasan S, Kumarasamy S, Andreadakis ZE, Lind PG (2023) Artificial intelligence and mathematical models of power grids driven by renewable energy sources: a survey. Energies 16(14):1–56. https://doi.org/10.3390/en16145383
63. Yang P (2014) Power system state estimation and renewable energy optimization in smart grids. Washington University in St, Louis
64. Chi Y, Tang B, Hu J, Member S, Tian X, Tang H, Li Y, Sun S (2019) Overview of mechanism and mitigation measures on multi-frequency oscillation caused by large-scale integration of wind power. CSEE J Power Energy Systems 5(4):433–443. https://doi.org/10.17775/cseejpes.2019.01100
65. Xu Q, Cao Y, Zhang H, Zhang W, Terzija V (2021) Bi-level dispatch and control architecture for power system in China based on grid-friendly virtual power plant. Appl Sci (Switz) 11(3):1–19. https://doi.org/10.3390/app11031282
66. Caramanis M, Ntakou E, Hogan WW, Chakrabortty A, Schoene J (2016) Co-optimization of power and reserves in dynamic T&D power markets with nondispatchable renewable generation and distributed energy resources. Proc IEEE 104(4):807–836. https://doi.org/10.1109/JPROC.2016.2520758
67. Shaikh JA, Mirjat NH, Memon ZA, Shan SA, Shaikh SA (2021) Optimization of renewable energy based microgrid for Mehran UET Jamshoro. J Phys Conf Ser 1860(1):01. https://doi.org/10.1088/1742-6596/1860/1/012022
68. Tang Z, Yang Y, Blaabjerg F (2022) Power electronics: the enabling technology for renewable energy integration. CSEE J Power Energy Syst 8(1):39–52. https://doi.org/10.17775/CSEEJPES.2021.02850
69. Iweh CD, Gyamfi S, Tanyi E, Effah-Donyina E (2021) Distributed generation and renewable energy integration into the grid: prerequisites, push factors, practical options, issues and merits. Energies 14(17):5375
70. Ali HH, Kassem AM, Al-Dhaifallah M, Fathy A (2020) Multi-verse optimizer for model predictive load frequency control of hybrid multi-interconnected plants comprising renewable energy. IEEE Access 8:114623–114642. https://doi.org/10.1109/ACCESS.2020.3004299
71. Meshram A (2022) Power quality improvement methodology in microgrid: a review. Int Res J Engg Technol (IRJET) 9(11):480–488
72. Fischer D, Madani H (2017) On heat pumps in smart grids: a review. Renew Sust Energ Rev 70:342–357. https://doi.org/10.1016/j.rser.2016.11.182
73. Krechel T, Sanchez F, Gonzalez-Longatt F, Chamorro HR, Rueda JL (2019a) A transmission system friendly micro-grid: optimising active power losses. In: 2019 IEEE Milan PowerTech. PowerTech, pp 1–6. https://doi.org/10.1109/PTC.2019.8810894
74. Şerban IOAN, Marinescu C (2008) Power quality issues in a stand-alone microgrid based on renewable energy. Rev Roum Sci Techn–Électrotechn et Énerg 53(3):285–293
75. Bayindir R, Colak I, Fulli G, Demirtas K (2016) Smart grid technologies and applications. Renew Sust Energ Rev 66:499–516. https://doi.org/10.1016/j.rser.2016.08.002
76. Sahoo SK (2016) Renewable and sustainable energy reviews solar photovoltaic energy progress in India: a review. Renew Sust Energ Rev 59:927–939. https://doi.org/10.1016/j.rser.2016.01.049
77. Kumar V, Pandey AS, Sinha SK (2016) Grid integration and power quality issues of wind and solar energy system: a review. In: International Conference on Emerging Trends in Electrical, Electronics and Sustainable Energy Systems, ICETEESES 2016, 2011, pp 71–80. https://doi.org/10.1109/ICETEESES.2016.7581355

78. Yong JY, Ramachandaramurthy VK, Tan KM, Mithulananthan N (2015) A review on the state-of-the-art technologies of electric vehicle, its impacts and prospects. Renew Sust Energ Rev 49:365–385. https://doi.org/10.1016/j.rser.2015.04.130
79. Zhang H, Xiang W, Lin W, Wen J (2021) Grid forming converters in renewable energy sources dominated power grid: control strategy, stability, application, and challenges. J Mod Power Syst Clean Energy 9(6):1239–1256. https://doi.org/10.35833/MPCE.2021.000257
80. Raut US, Yee DS (2017) Optimization of microgrid for renewable power integration. IARJSET 4(2):61–64. https://doi.org/10.17148/iarjset.2017.4213
81. Bade SK, Kulkarni VA (2018) Analysis of railway traction power system using renewable energy: a review. In: 7th IEEE International Conference on Computation of Power, Energy, Information and Communication. ICCPEIC, pp 404–408. https://doi.org/10.1109/ICCPEIC.2018.8525206
82. Sandhu M, Thakur T (2014) Issues, challenges, causes, impacts and utilization of renewable energy sources - grid integration. J Engg Res Appl 4(3):636–643
83. Vitharanage GD, Vilathgamuwa M, Mishra Y, Cory P (2023, November) Optimal scheduling of electric vehicles in vehicle-to-grid system with battery energy storage for cost-efficient integration of distributed photovoltaics. In: 2023 IEEE 8th Southern Power Electronics Conference and 17th Brazilian Power Electronics Conference (SPEC/COBEP). IEEE, pp 1–8
84. Zeng M, Duan J, Wang L, Zhang Y, Xue S (2015) Orderly grid connection of renewable energy generation in China: management mode, existing problems and solutions. Renew Sust Energ Rev 41:14–28. https://doi.org/10.1016/j.rser.2014.08.047
85. Olulope PK, Folly KA, Venayagamoorthy GK (2013, February) Modeling and simulation of hybrid distributed generation and its impact on transient stability of power system. In: 2013 IEEE International Conference on Industrial Technology (ICIT). IEEE, pp 1757–1762
86. Ahsan F, Dana NH, Sarker SK, Li L, Muyeen SM, Ali MF, Tasneem Z, Hasan MM, Abhi SH, Islam MR, Ahamed MH, Islam MM, Das SK, Badal MFR, Das P (2023) Data-driven next-generation smart grid towards sustainable energy evolution: techniques and technology review. Prot Control Mod Power Syst 8(1):43. https://doi.org/10.1186/s41601-023-00319-5
87. Babonneau F, Caramanis M, Haurie A (2016) A linear programming model for power distribution with demand response and variable renewable energy. Appl Energy 181:83–95. https://doi.org/10.1016/j.apenergy.2016.08.028
88. Behboodi S, Chassin DP, Crawford C, Djilali N (2016) Electric vehicle participation in transactive power systems using real-time retail prices. In Proceedings of the Annual Hawaii International Conference on System Sciences, 2016-March, 2400–2407. https://doi.org/10.1109/HICSS.2016.300
89. Adefarati T, Bansal RC (2016) Integration of renewable distributed generators into the distribution system: a review. IET Renew Power Gen 10(7):873–884. https://doi.org/10.1049/iet-rpg.2015.0378
90. Roy P, He J, Zhao T, Singh YV (2022) Recent advances of wind-solar hybrid renewable energy systems for power generation: a review. IEEE Open J Indus Electron Soc 3:81–104. https://doi.org/10.1109/OJIES.2022.3144093
91. Agarwala A, Tahsin T, Ali MF, Sarker SK, Abhi SH, Das SK, Hasan M, Zinat Tasneem M, Islam MB, Ahamed MH (2024) Towards next generation power grid transformer for renewables: technology review. Eng Rep 6(4):e12848
92. Ramesh J, Shahriar S, Al-Ali AR, Osman A, Shaaban MF (2022) Machine learning approach for smart distribution transformers load monitoring and management system. Energies 15 (21):1–19. https://doi.org/10.3390/en15217981
93. Ranjan S, Jaiswal S, Latif A, Das DC, Sinha N, Hussain SMS, Ustun TS (2021) Isolated and interconnected multi-area hybrid power systems: a review on control strategies. Energies 14 (24):8276. https://doi.org/10.3390/en14248276
94. Cabrera-Tobar A, Bullich-Massagué E, Aragüés-Peñalba M, Gomis-Bellmunt O (2016) Topologies for large scale photovoltaic power plants. Renew Sust Energ Rev 59:309–319. https://doi.org/10.1016/j.rser.2015.12.362

95. Bahta ST (2013) Design and analyzing of an off-grid hybrid renewable energy system to supply electricity for rural areas. KTH School of Industrial Engineering and Management
96. Rakhshani E, Rouzbehi K, Sánchez J, Tobar AC, Pouresmaeil E (2019) Integration of large scale PV-based generation into power systems: a survey. Energies 12(8):1425
97. Bajaj M, Singh AK (2020) Grid integrated renewable DG systems: a review of power quality challenges and state-of-the-art mitigation techniques. Int J Energy Res 44(1):26–69. https://doi.org/10.1002/er.4847
98. Nghitevelekwa K, Bansal RC (2018) A review of generation dispatch with large-scale photovoltaic systems. Renew Sust Energ Rev 81:615–624. https://doi.org/10.1016/j.rser.2017.08.035
99. Cecati C, Citro C, Siano P (2011) Combined operations of renewable energy systems and responsive demand in a smart grid. IEEE Trans Sustain Energy 2(4):468–476. https://doi.org/10.1109/TSTE.2011.2161624
100. Dileep G (2020) A survey on smart grid technologies and applications. Renew Energy 146:2589–2625. https://doi.org/10.1016/j.renene.2019.08.092
101. García-López FP, Barragán-Villarejo M, Maza-Ortega JM (2020) Grid-friendly integration of electric vehicle fast charging station based on multiterminal DC link. Int J Electr Power Energy Syst 114:105341. https://doi.org/10.1016/j.ijepes.2019.05.078
102. Avancini DB, Rodrigues JJPC, Martins SGB, Rabêlo RAL, Al-Muhtadi J, Solic P (2019) Energy meters evolution in smart grids: a review. J Clean Prod 217:702–715. https://doi.org/10.1016/j.jclepro.2019.01.229
103. Kohlhepp P, Harb H, Wolisz H, Waczowicz S, Müller D, Hagenmeyer V (2019) Large-scale grid integration of residential thermal energy storages as demandside flexibility resource: a review of international field studies. Renew Sust Energ Rev 101:527–547. https://doi.org/10.1016/j.rser.2018.09.045
104. Faheem M, Shah SBH, Butt RA, Raza B, Anwar M, Ashraf MW, Ngadi MA, Gungor VC (2018) Smart grid communication and information technologies in the perspective of industry 4.0: opportunities and challenges. Comput Sci Rev 30:1–30. https://doi.org/10.1016/j.cosrev.2018.08.001
105. Kenyon RW, Bossart M, Marković M, Doubleday K, Matsuda-Dunn R, Mitova S, Julien SA, Hale ET, Hodge BM (2020) Stability and control of power systems with high penetrations of inverter-based resources: an accessible review of current knowledge and open questions. Sol Energy 210:149–168
106. Ourahou M, Ayrir W, Hassouni ELB, Haddi A (2020) Review on smart grid control and reliability in presence of renewable energies: challenges and prospects. Math Comput Simul 167:19–31. https://doi.org/10.1016/j.matcom.2018.11.009

Parameter Identification of Empirical Models for Head Estimation of Electrical Submersible Pumps used for Dual-phase Petroleum Fluids

Saud Al-Ghaithi, Morteza Mohammadzaheri, and Ali Al-Humairi

Abstract The performance of electrical submersible pumps (ESPs) in petroleum fluids decreases significantly when operating under dual-phase (gas–liquid) flow conditions compared to single-phase operations. Two-phase flow conditions also creates difficulties in head prediction, which is a vital for petroleum production system design and operation. This chapter solves the problem by evaluating and validating empirical head prediction models and identifying their new parameters for two-phase flow situations. The study examines multiple existing head-prediction models and optimize their parameters to enhance their ability to model gas–liquid mixture behavior. The research combines empirical modeling with statistical parameter identification through the application of Ordinary Least Squares (OLS) fitting to extensive experimental data and data filtering methods to reduce measurement noise and achieve robust parameter estimation.

The research demonstrates thorough systematic model calibration and validation so that the adjusted models better predict head than uncalibrated models while providing reliable head predictions across various gas volume fractions without requiring complex analytical methods. The practical value of this research becomes evident through a user-friendly graphical user interface (GUI) that uses the calibrated models to let engineers easily calculate ESP head under different two-phase flow conditions. The research findings have been transformed into a decision-support tool by the (GUI) which demonstrates the methodological strength of this chapter while providing a practical solution that can be enhanced for various field applications.

Keywords Electrical submersible pumps · Empirical models · Multiphase petroleum fluid · Petroleum Industries · Artificial Lift

---

S. Al-Ghaithi · M. Mohammadzaheri (✉) · A. Al-Humairi
German University of Technology in Oman (GUtech), Halban, Oman
e-mail: morteza.zaheri@gutech.edu.om

© The Author(s), under exclusive license to Springer Nature Singapore Pte Ltd. 2026

A. Azizi (ed.), *Hybrid Energy Systems and AI Integration*, Emerging Trends in Mechatronics, https://doi.org/10.1007/978-981-95-5859-9_3

Abbreviations

Symbol/Abbreviation	Unit/Full Form
ANN	Artificial neural network
CO_2	Carbon dioxide
CV	Cross-validation
ESP	Electric submersible pump
FCC	Fully-connected cascade
GPM	Gallons per minute
GUI	Graphical user interface
GVF	Gas volume fraction
IPR	Inflow-performance
MAE	Mean Absolute Error
OLS	Ordinary least squares
psig	Gage pressure (pounds per square inch)
RBF	Radial basis function
RMSE	Root Mean Squared Error
rpm	Revolutions per minute
VSD	Variable-speed drives
H_m	Head of mixture (ft)
H_l	Head of liquid (ft)
x_1,x_2	Turpin [1] fitted coefficients
P_{in}	Inlet pressure (psi)
φ	Phi, Turpin [1] surging limit
q	Mixture flow rate (GPM)
q_l	Liquid flow rate
q_g	Gas flow rate
γ_{mix}	Gamma, mixture density
θ	vector of coefficients

1 Introduction and Literature Review

1.1 Background

As Weriono et al. [2] stated, "Electric Submersible Pumps (ESPs) are one of the most prominent artificial lifting methods in the oil and gas industry today" (p. 107), as they excel at handling high-volume flow rates [3]. ESPs' use in wells with gas-liquid mixtures has created performance degradation problems which continue to drive engineering research [4]. The intake fluid becomes less efficient when free gas exists in the fluid because it decreases the generated head, while leading to gas locking and surging instabilities.

Operational challenges of two-phase flow in ESPs have led engineers and researchers to create different mathematical models that forecast their performance, i.e. head. The available models consist of mechanistic methods that solve conservation equations alongside empirical correlations which use experimental data to develop mathematical relationships. Empirical models persist as important tools despite numerical progress because they provide easy implementation with quick computations and simple field-level diagnostics. In addition, more advanced artificial neural networks and fuzzy logic models have been introduced that trade interoperability with very high accuracy [5, 6, 26].

1.2 Problem Statement

Field engineers together with production optimization teams need precise calculations of ESP head performance under two-phase gas-liquid petroleum fluid operations. Existing empirical models fail to perform well when used outside the operating areas where parameters were established during their original development process. The models currently in use do not properly address real-world constraints because they fail to set boundaries based on single-phase head restrictions or the onset of instability at high gas volume fractions [7]. The evaluation process of empirical models on actual data needs improvement especially using parameter identification methods and filtering procedures.

Some empirical correlations get deployed without proper evaluation of their regression domain or filtering criteria [8]. The absence of methodological rigor in modeling leads to results that either match the training data too closely or produce unrealistic patterns [9]. The development of this research serves as a structured method which includes both model calibration with new experimental data along with structured data filtering techniques to enhance predictive reliability and a comparison of applied models.

1.3 Research Objectives

The main objective of this chapter is the evaluation and calibration process of empirical models which predict the head output of electrical submersible pumps (ESPs) under petroleum gas-liquid flow situations. The research examines three prominent models which are Turpin [1], Sachdeva [10], and Zhou and Sachdeva [4]. The chapter establishes a MATLAB-based live script to perform comprehensive model evaluation and improvement.

The study executes an interpretable parameter identification method through ordinary least squares (OLS) regression analysis. Transformation methods were used to convert nonlinear models into linear ones to achieve transparent mathematical solutions. Each model's accuracy and robustness get evaluated using Mean

Absolute Error (MAE) and Root Mean Squared Error (RMSE) performance metrics which enable evaluation of prediction accuracy across different operating conditions.

The research introduces organized data filtering approaches as a fundamental aspect of its work. A novel approach developed in this work, head-comparison filter, eliminates data points that show unrealistic physical behavior and mainly affects Sachdeva model predictions. The regression results become more interpretable and consistent due to this filtering process.

This research employs visual diagnostics through predicted-versus-measured plots and residual distributions and performance curve overlays to determine how well each model matches observed data. The final model selection process leads to the implementation of the Zhou and Sachdeva model within a MATLAB App Designer-based user-friendly graphical user interface (GUI). The practical tool enables field engineers to estimate ESP performance rapidly under various conditions through a user-friendly interface which does not need extensive modeling knowledge.

1.4 Research Scope

This research focuses exclusively on three existing empirical models used for ESP head prediction in multiphase petroleum flow operations. The analysis evaluates how accurately these three models match experimental data for CO_2/diesel gas-liquid mixtures across different flow velocities and inlet pressure conditions. The authors transform each model into linear form (where necessary) then perform log-transformed regression analysis to validate predictions against experimental measurements including unfiltered and filtered data.

This research does not explore mechanistic or neural network models because the goal is to preserve interpretability together with minimal computational requirements and practical field deployment capabilities. The proposed graphical user interface tool was built with MATLAB App Designer to deliver instant field-side pump performance estimates through operator input [11].

1.5 Methodological Overview

The research methodology within this chapter uses a systematic approach to evaluate and validate empirical head prediction models for ESP operations in dual-phase petroleum flow systems. The methodology includes multiple structured steps which follow strict implementation procedures to maintain both reproducibility and physical system compatibility. A complete predictive framework emerges from this systematic process which achieves both scientific validity and practical applicability.

The initial step involves data preprocessing based on the data presented in Lea and Bearden [12]. Data is imported to MATLAB as complete experimental dataset containing inlet pressure data alongside gas volume fraction measurements, flow rate data and measured pump head readings. The dataset from controlled CO_2 and diesel mixtures serves as the empirical basis for evaluating and calibrating the models. The data processing at this stage maintains its original characteristics because no smoothing or pre-cleaning techniques are applied.

Each model receives specialized filtering techniques which remove data points that show physical inconsistencies or unstable behavior. The Turpin model uses the surge indicator $\varphi < 1$ to prevent calibration from occurring in unstable flow conditions according to Turpin et al. [1]. The Sachdeva correlation uses a novel filtering tool which compares measured two-phase head with single-phase liquid head obtained from a Simulink model developed for this purpose, presented in the Appendix. The model retains data points that show measured head values below the liquid-only head values to enhance the physical accuracy of regression results. The Zhou and Sachdeva [4] model employs critical flow filtering to exclude points which indicate choked flow through a recognized flow ratio criterion.

Model generalizability was evaluated using a train-test split approach, where 80% of the available data was used for training and 20% was kept for testing. This validation method was applied to the Sachdeva [10] pressure-rise model and the Zhou and Sachdeva [4] model to have an insight of predictive performance under realistic conditions. The 80/20 split was chosen to ensure that an enough amount of data remained available for parameter estimation while still providing an independent test set to assess model accuracy.

For the Turpin [1] model, the full filtered dataset was used directly for parameter estimation without splitting. This choice was due to the limited number of valid data points remaining after applying the surging criteria. Performing a data split in this case would have reduced the training sample.

After data filtration the empirical models need logarithmic reformulation when necessary before linear regression can be applied. The transformation of data through ordinary least squares (OLS) fitting produces analytical solutions for the model parameters in the power-law and exponential formulations of the models respectively [13]. The complete MATLAB implementation uses matrix algebra for these regression operations which makes them transparent and compatible for student and field environments that lack advanced toolboxes.

The performance evaluation uses two error metrics which are Mean Absolute Error (MAE) and Root Mean Squared Error (RMSE). The two evaluation metrics demonstrate prediction accuracy differently because MAE shows average deviation while RMSE detects outliers. Each model's predictions receive evaluation through direct benchmarking against the recorded head values to establish performance comparison.

Visual tools supplement the numerical analysis. The alignment of predicted values to observed behavior appears in predicted-versus-measured plots while residual plots demonstrate systematic error patterns and bias. Performance curves containing model overlays are available for specific operating points to evaluate the models' ability to mimic real-world trends including head degradation as flow rate increases.

The developed graphical user interface (GUI) of MATLAB App Designer implements the model with the most consistent performance as we will observe further on which is Zhou and Sachdeva [4]. Through this interface users can provide operating conditions to obtain real-time predicted head values. The developed GUI serves as an operational tool which technicians can use to implement the research findings in their work environment.

The implementation of each step remains consistent throughout all model applications to filtered datasets which enables fair model comparison while preserving the structural differences between models. The end-to-end approach to methodology creates both model behavior validation and final implementation interpretability and field-readiness.

1.6 Significance of the Study

The research combines a systematic filtering approach with strong parameter estimation and performance assessment to develop a reproducible field-tested method for ESP head model evaluation. The project uses real experimental data to maintain practicality while providing a GUI interface that meets industry requirements for deployable tools.

The novel head-comparison mask functions as a unique filtering strategy which adds academic and operational value to the research. The approach shows how domain knowledge enables improved realism and robustness of empirical correlations.

1.7 Structure of this Chapter

This chapter follows a five-section structure which progresses systematically from fundamental context to practical implementation while maintaining logical continuity between sections. This section reviews the existing literature about dual-phase petroleum flow head-estimation methods including empirical, mechanistic and hybrid approaches for electrical submersible pumps (ESPs). The academic basis for re-examining and updating these three models [1, 4, 10] receives special focus in this work.

The third section explains the research methodology by describing all steps from experimental data import through data preparation to filtering and model transformation and parameter fitting with ordinary least squares regression and model evaluation using numerical and visual assessment tools. The research methodology follows a structured approach which guarantees both reproducibility and transparency and maintains compliance with physical restrictions found in ESP systems operating under multiphase flow.

Section 4 presents the core results and analysis. The assessment of model performance relies on statistical metrics including Mean Absolute Error (MAE) and Root Mean Squared Error (RMSE) together with predicted-versus-measured comparisons and residual plots and performance curve overlays. The Sachdeva correlation experiences improved predictive validity because of the novel filtering strategy implemented in this research. The analysis in this section leads to the choice of Zhou & Sachdeva as the model for field deployment through a graphical user interface (GUI).

The research study concludes in Sect. 4 with a summary of its main findings. The study presents essential contributions alongside deployment limitations and future research directions for extending ESP modeling in multiphase environments and field deployment implications.

This study presents a technically grounded evaluation which provides practical solutions to model ESP performance during petroleum two-phase flow operations.

1.8 Fundamentals of Electrical Submersible Pump

1.8.1 Introduction and History of ESPs

Arutunoff patented the electrical submersible pump (ESP) in 1926 as the first device to enhance deep well liquid extraction [14]. The early designs of ESPs underwent multiple developments since the 1950s until the 1970s when seal "protectors" and rotary gas separators and variable-speed drives VSD were added for improved reliability and gas handling capabilities [15].

1.8.2 A Basic ESP System

The lowest part of the ESP—See Fig. 1—string holds a three-phase induction motor. The downhole motor transforms electrical power into mechanical rotation which operates the pump stages. The wellbore supports a three-conductor cable that connects surface switchgear to the motor [3].

The protector section acts as a seal for the motor housing while it holds the thrust bearings directly above the motor. The protector maintains motor shaft reliability by supporting axial loads and isolating motor oil under high downhole pressures [3].

Following the motor there is a gas-separator (or intake) stage. The fluid entering the ESP reaches a velocity that enables free gas to separate from the fluid before it rises back to the wellbore. The removal of gas before it enters the primary pump sections stops gas locking and enables steady pump operation [3].

Above the separator is the multistage centrifugal pump itself. A single stage of this pump consists of an impeller that produces fluid acceleration and a diffuser

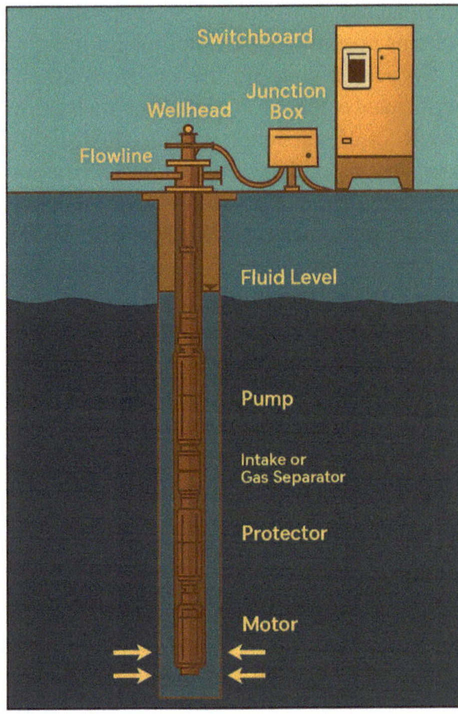

Fig. 1 Basic ESP assembly

which transforms the fluid velocity into pressure. The pump reaches its required total head by placing multiple impeller–diffuser pairs on top of each other.

The whole assembly descends into the well through production tubing while the power cable runs parallel to it. The well fluids move through the motor cooling section before entering the pump stages where they produce the mixture that flows up through the tubing and wellhead [3].

1.8.3 Working Principles of an ESP

An ESP stage contains two essential components which are impeller and diffuser. The spinning impeller generates fluid velocity that the diffuser transforms into pressure. The total head H results from adding together all individual stage heads at a specific flow rate q. The motor operates as a Y-connected squirrel-cage induction design which operates near synchronous speed (e.g., 3500 rpm on 60 Hz). The motor uses circulating motor oil through hollow shafts to keep the windings and bearings cool [16].

1.8.4 Deploying an ESP System

The pump performance requires matching with well inflow by ensuring the pump's head–flow curve corresponds with the reservoir's inflow-performance (IPR) curve.

The motor starts by drawing 4–6 times its operating current from surface power until it reaches its stable running range. The installation of variable-speed drives VSD reduces the high starting currents by controlling the voltage sent to the motor but requires additional equipment along with increased system complexity [15]

1.9 Dual-Phase Operation: Effects of Gas on ESP Performance

The flow characteristics and head-flow performance and efficiency of ESPs change when free gas enters the intake fluid because the two-phase mixture shows different compressibility and density properties than single-phase liquid. The initial research demonstrated that any amount of gas in the liquid decreases pump operation yet creates separate operational zones based on increasing gas void fraction (GVF) [12].

At low GVF—typically below 3%–5%—bubbles remain small and dispersed, following the liquid streamlines through the impeller. Under these "homogeneous" conditions, head and efficiency decline only slightly relative to liquid-only operation, and the pump curve retains its shape with minor attenuation [12, 17].

When the gas void fraction reaches between 5% and 7%, bubbles start to merge into separate gas pockets which form near the impeller eye and shroud area. The pump head experiences periodic low-frequency oscillations known as surging during this transition phase because the pump cycles between liquid-rich and gas-rich slug ingestion which results in substantial deviations from the liquid-only head–capacity curve [10, 18]

The pump loses its ability to maintain pressure rise when gas void fraction exceeds 10%. The pump experiences a complete breakdown of its head during this "gas locking" regime until the liquid supply to the impeller eye becomes sufficient to prevent total shutdown [17, 18].

The first-order effect of intake pressure determines these thresholds. At elevated suction pressure levels the compressed bubbles take longer to form pockets which extends the operational range to ~15 % GVF at 200–300 psig in diesel–CO_2 tests compared to <5 % at near-atmospheric conditions [10]

The correlations developed by researchers provide stage-by-stage head loss predictions based on gas void fraction and inlet pressure measurements. [19] developed stage-by-stage correction methods which enable users to predict head loss deterioration and [20] evaluated surging correlations which relate GVF thresholds to pump design parameters and operational settings.

Studies like Zhu et al. [18] show that ESP operation depends heavily on GVF control and inlet pressure management to achieve reliable performance in wells with gas content. Most predictive models function under constant impeller speed conditions while simplifying bubble behavior to address transitional operating ranges that are relevant to real-world conditions.

1.10 Overview of Modeling Approaches

1.10.1 Analytical Models

The fundamental mass and momentum balances of impeller–diffuser stages form the basis of analytical modeling. According to Nguyen [3], the methods use simplified one-dimensional flow equations to derive idealized gas correction terms which stem from homogeneous mixture assumptions. The models provide valuable physical understanding, but their basic simplifications restrict their ability to produce precise results for different two-phase operational conditions [21].

1.10.2 Empirical Models

Experimental two-phase pump data receives its form through the application of empirical correlations that establish closed-form equations. The head correction model of Turpin [1] utilized three variables including inlet pressure and gas void fraction and flow rate exponents to predict diesel–CO_2 mixture head loss at constant speed [1]. The research of Sachdeva et al. [10] created correlational factors which stage-by-stage calculated head-capacity relationships for axial and radial ESP configurations to provide simple head-capacity predictions under gassy flow conditions [10].

1.10.3 Neural Network Models

Neural networks function as data-driven systems which translate operating parameters into pump head output. The shallow feed-forward ANN proposed by Mohammadzaheri et al. [6] demonstrated a 69% improvement in mean absolute error when trained with diesel–CO_2 test data [6]. Mohammadzaheri et al. [26] presented radial basis function RBF networks that used optimized spread parameters and cross-validation across 110 experiments to achieve robust generalization throughout the entire operational range [5]. Neural network models are not used in this chapter due to their black-box nature and high data requirements, which limit interpretability and practical deployment in field environments.

1.10.4 Fuzzy Logic Models

The Sugeno-type fuzzy inference systems apply rule-based methods to handle the uncertainties present in two-phase systems. A Sugeno-type model developed by Mohammadzaheri et al. [5] used Gaussian membership functions for inlet pressure and flow rate and gas void fraction to build weighted rules for head estimation with an absolute error that reduced by 52.4% compared to empirical correlations

Table 1 Comparison of modeling approaches

	Interpretability	Data requirements	Computational complexity
Analytical models	High – based on clear physical laws	Low – minimal test data needed	Low – straightforward calculations
Empirical models	Moderate – tied to empirical fits	Moderate – requires two-phase pump experiments	Moderate – parameter estimation via curve fitting
Neural network models	Low – black-box weight-based	High – large datasets for training and validation	High – iterative training and tuning
Fuzzy logic models	Moderate – rule-based transparency	High – expert knowledge and training samples	Moderate – rule evaluation and defuzzification

[5]. Fuzzy logic systems, despite their accuracy in handling uncertainty, are excluded from this study to maintain methodological simplicity and ensure compatibility with conventional engineering workflows.

1.10.5 Section Conclusion

Analytical models establish basic principles yet produce insufficient practical results while neural and fuzzy methods provide high accuracy but demand complex models and extensive data requirements. Empirical correlations demonstrate the best combination of interpretability along with computational simplicity and diesel–CO_2 prediction accuracy therefore they serve as the best starting point for this chapter (Table 1).

1.11 Empirical Head-Estimation Models

1.11.1 Role of Empirical Models

Empirical correlations enable direct head estimation of ESP through mathematical relationships between measurable parameters such as inlet pressure and gas void fraction and flow rate that result in performance predictions. The analytical models solve complex two-phase interactions without needing to solve detailed flow equations. Their basic nature makes them suitable for quick calculations as well as data adjustment and field measurements and design applications [1, 4, 10].

1.11.2 Turpin [1] Model

The Turpin model predicts head loss in diesel–CO_2 mixtures at 3500 rpm by using power-law expressions which combine inlet pressure and gas void fraction and flow rate measurements. The segmented curve fitting method produced the coefficients by assuming homogeneous bubble dispersion while ignoring slip velocity between phases. The model provides reliable predictions of head reduction throughout low- to moderate GVF ranges yet produces less accurate results in conditions where discrete gas pockets form [1].

$$\frac{H_m}{H_l} = exp\left(x_1\left(\frac{q_g}{p_{in}q_l}\right)^2 + x_2\left(\frac{q_g}{p_{in}q_l}\right)\right), \tag{1}$$

$\frac{H_m}{H_l}$ = ratio of multiphase to liquid-only head. q_g, q_l = gas & liquid flow rates. p_{in} = inlet pressure. x_1, x_2 = fitted coefficients.

The pump enters a surging state when the parameter ϕ exceeds 1 according to Turpin [1]. The surging criteria is stated below,

$$\phi_{Turpin} = \left(\frac{2000}{3p_{in}}\right)\frac{q_g}{q_l} \tag{2}$$

Note that the liquid and gas flow rate can be calculated by the following equation:

$$\alpha = \frac{q_g}{q_g + q_l}, \tag{3}$$

where α is the gas void fraction (GVF).

1.11.3 Sachdeva et al. [10]

Sachdeva et al. created a stage-by-stage multiplicative relationship to calculate the head per stage through power-law functions of pressure and gas fraction and flow. The formula functions with various ESP geometries and allows modification for different stage configurations. Although it matches well with moderate GVF data the authors did not establish a specific threshold for surging or gas-locking behavior when flow approaches critical conditions [10]

$$\Delta p_m = K_2\, p_{in}^{E_1} \alpha_G^{E_2} q_l^{E_3}, \tag{4}$$

α = gas void fraction. E_1, E_1, E_1 and K_2 = models' parameters.

Surging criterion: none defined in the original publication.

1.11.4 Zhou and Sachdeva [4]

The authors Zhou and Sachdeva developed a hybrid correlation by combining the manufacturer's single-phase head curve with gas correction exponents and a unitary

Table 2 Literature parameters [4]

Pump type	Model	K_C	E_4	E_5
Radial	I-42B	1.418884	−0.07244	0.318544
Radial	0–72	2.173723	−0.05253	0.833587
Mixed	K-70	7.497750	−0.34870	0.867092

Table 3 Surging correlations and critical conditions

Model (Year)	Surge/Critical Condition	Notes
Turpin [1]	$\phi_{Turpin} = \left(\frac{2000}{3p_{in}}\right)\frac{q_g}{q_l}$	Dimensionless parameter defining the onset of surging regime.
Sachdeva et al. [10]	None	No explicit surge or stall criterion provided [10]
Zhou and Sachdeva [4]	$(q_m)_{max} = K_c\, q_{max}\, {p_{in}}^{E_7}\, \alpha_G^{E_8}$	Maximum allowable flow enforcing the surge boundary.

factor (C). The approach preserves catalog head–flow behavior under no-gas conditions before smoothly moving to two-phase operation. The correction factor (K) needs re-evaluation for different pump models since the predictions follow the specifications. Given below is the empirical model in its original nonlinear form [4]

$$\frac{H_m}{H_{max}} = K(C_1 p_{in})^{\alpha E_1}\,(1-\alpha)^{E_2}\left(1-\frac{q_m}{q_{max}}\right)^{E_3}, \tag{5}$$

The model includes a total flow limit to determine the surge boundary expressed as:

$$(q_m)_{max} = K_c\, q_{max}\, {p_{in}}^{E_7}\, \alpha_G^{E_8}, \tag{6}$$

where E_7, E_8, and K_c are obtained from the literature as follows (Table 2):

1.11.5 Model Limitations

These empirical correlations need new two-phase dataset-based coefficient optimization for each application since their original parameters lack universality according to studies such as Ali et al. [9] between different fluids and operating conditions. Each model stops giving accurate head performance predictions when its specified surging or gas-locking threshold is exceeded (Table 3).

1.12 *Parameter Identification Techniques and Performance Metrics*

1.12.1 Ordinary Least Squares (OLS)

The main method of OLS regression uses the least squares method to determine empirical model coefficients by finding the minimum sum of squared differences between estimated and actual head values [22]. The process of linearization through logarithmic transformation allows OLS to directly solve normal equations for model equations that contain nonlinear parameters. The method provides unbiased parameter estimates when errors show constant variance and independence while enabling residual diagnostics for detecting model inadequacies. The robustness of OLS remains high for small data sets because it uses all available data without partitioning which results in maximum estimation reliability [23].

1.12.2 Mean Absolute Error (MAE)

The Mean Absolute Error (MAE) evaluates model accuracy through the calculation of average absolute differences between predicted and observed values. MAE evaluates all deviations at a linear level which reduces outlier sensitivity while providing straightforward interpretation in the target variable units. The widespread use of MAE in two-phase ESP studies demonstrates its effectiveness for evaluating model bias together with typical deviations under field tolerance conditions [23].

1.12.3 Root Mean Squared Error (RMSE)

Root Mean Squared Error (RMSE) is used as a supplementary measure of accuracy which computes the square root of the mean of squares of the difference between the predicted and actual values. In contrast to MAE, RMSE will put more weight to greater deviations since it squares the values hence, this type of measure is especially sensitive to outliers and gross under- or over prediction. The sensitivity enables RMSE to point at a region where the model performs adversely, providing information into the variance-based deficiencies. The main reason RMSE is still popular with engineering applications is since it indicates the amount and distribution of the predictive errors in a one-dimensional figure which thus it is able to compare models under different operating conditions [23].

1.13 *Summary and Research Gaps*

This research investigates the two-phase ESP head estimation using different modeling techniques that present distinct advantages and weaknesses. Three empirical correlations by Turpin [1], Sachdeva et al. [10], and Zhou and Sachdeva [4] have developed direct and closed-form relationships between key operational variables (inlet pressure, gas void fraction, and flow rate) and pump head. The Turpin power-law formulation continues to be popular because it is simple to use yet Sachdeva's stage-wise correlation allows pump head prediction with geometric variations in axial and radial pump designs. The Zhou and Sachdeva hybrid model combines vendor catalog curves with gas correction exponents to extend applicability to commercial pump data. The head predictions from empirical methods demonstrate reliable accuracy through diesel–CO_2 test datasets because they maintain mean absolute errors below 10 % in moderate GVF conditions.

The literature lacks comparative studies between models and the current research fails to evaluate their performance using a single error metric. The literature shows individual reports with MAE or R^2 values for particular models but no study has evaluated the performance of all three empirical correlations across the same dataset. The lack of this comparison prevents practitioners from obtaining proper guidance regarding the most accurate prediction framework for different two-phase operating conditions.

Surging criteria exist explicitly for Turpin and Zhou and Sachdeva models through dimensionless parameters and maximum allowable flow limits that indicate operational instability. The 1992 Sachdeva correlation lacks defined surge thresholds which makes it difficult for users to determine when gas locking or head collapse begins. The absence of standard surge boundary evaluations across empirical formulations requires immediate attention.

All three models receive their coefficients through manual or segmented fitting methods which use specific diesel–CO_2 experiments. The models' parameter values show restricted generalizability to other dual-phase petroleum mixtures or extended GVF ranges. The main methodological weakness stems from the lack of unified data-driven calibration procedures that use ordinary least squares or nonlinear optimization methods across complete gas void fraction ranges. A solution to this challenge would unify coefficient estimation methods and boost model robustness for predictions that extend past original test parameters.

The current literature demonstrates effective empirical head prediction for ESP operations in gassy conditions yet it lacks essential comparisons between major correlations and unified surging criteria and full recalibration techniques. This chapter addresses these gaps through an evaluation of Turpin [1], Sachdeva [10], and Zhou and Sachdeva [4] correlations on a single diesel–CO_2 dataset at fixed 3500 rpm while using consistent mean absolute error assessment and defining unified surge criteria where absent. The research will establish precise guidelines for choosing empirical models together with their parameter estimation strategies in dual-phase ESP operations.

2 Methodology

2.1 Methodology Overview

The section demonstrates a complete method which applies to parameter identification and model evaluation of three head-estimation models designed for electrical submersible pumps operating under gas–liquid two-phase flow conditions in petroleum fluids. The models under investigation are (1) the Turpin [1] empirical model, (2) the Sachdeva [10] pressure-rise model, and (3) the Zhou and Sachdeva [4] model. The MATLAB environment allows matrix-based regression operations to derive best-fit coefficient values which match experimental data points in each model.

The calibration process transforms nonlinear correlations into linear equations through natural logarithm operations applied to dependent and independent variables. The log-transformed datasets undergo ordinary least squares (OLS) regression which produces closed-form solutions for model coefficients while maintaining complete reproducibility and transparency this gives an advantage over nonlinear solvers which are not as transparent hence we use it. OLS solutions are not affected by the parameter values provided from literature for validation purposes.

The calibration process transforms nonlinear correlations into linear equations through natural logarithm operations applied to dependent and independent variables. The log-transformed datasets undergo ordinary least squares (OLS) regression which produces closed-form solutions for model coefficients while maintaining complete reproducibility and transparency this gives an advantage over nonlinear solvers which are not as transparent hence we use it. OLS solutions are not affected by the parameter values provided from literature for validation purposes.

The experimental data, used in this research, are fully presented in the appendix to support complete model calibration reproduction. The measurements from the CO_2/diesel two-phase mixture cover a broad range of inlet pressures and gas volume fractions that match typical ESP operating conditions.

The selection of specific data points for each empirical model happens through a purpose-designed filtering process. The Turpin model uses surge criterion ($\phi < 1$) to filter data points based on gas-to-liquid flow ratio compared to inlet pressure while eliminating unstable operational conditions. The Sachdeva model undergoes evaluation through two testing methods: its basic unfiltered version and a proposed head-comparison filter which deletes measurement points that show two-phase head exceeding liquid-only head values obtained from Simulink pump simulations. This additional mask refines the regression domain to realistic two-phase behavior. A critical flow ratio criterion from published coefficients is used by the Zhou and Sachdeva model to eliminate surging regimes before the fitting process begins [4].

Model performance evaluation relies on two error assessment methods first, the mean absolute error (MAE) and the root mean squared error (RMSE). The performance assessment through these metrics enables the comparison of model precision and stability between different operational conditions. The section finishes by presenting a visualization approach which combines experimental and predicted head values with flow data for specific inlet pressure and gas fraction combinations while using residual scatter plots to show error patterns for all model variants.

A MATLAB graphical user interface (GUI) was developed to support practical application and rapid scenario analysis. The GUI accepts user-defined inputs for inlet pressure (p_{in}), gas volume fraction (α), and mixture flow rate (q_m), and instantly computes the estimated head of the chosen model. giving the model equation and parameter set within an interactive environment thats can be simply used by technition and field engineers. All coding was performed in MATLAB Live Script, organized with clear headings and subheadings to serve as both execution code and a foundational guide for reproducibility and instructional purposes. The following sections explain data management methods and model equations and regression procedures and validation strategies as well as plotting techniques.

2.2 Data Import and Preprocessing

The chapter corresponding author delivered experimental measurements through a single MATLAB data file which the researcher imported in one operation to maintain all recorded operating points. The import script retrieves four essential variables which include inlet gage pressure (p_{in}), volumetric gas fraction (α), total mixture flow rate (q_m) and the resulting two-phase head (H). The Turpin [1] model uses the provided variables to calculate phase flow rates directly which results in liquid flow rate and gas flow rate. The import process does not include any initial smoothing or interpolation or outlier removal because the subsequent filtering will only result from model-specific criteria. Upon loading the data, the liquid and gas phase flow rates were computed by the following equations:

$$q_l = (1 - \alpha)\, q_m, q_g = (\,\alpha) q_m \tag{7}$$

The Simulink pump simulation produces single-phase (liquid-only) head (h_l) data at every experimental flow rate. The output from Simulink serves two purposes: it functions as a reference to evaluate mixture performance and it directly enters the Turpin model's exponential head correction. The direct relationship between experimental two-phase measurements and liquid-only head series is established through index-by-index alignment which enables the head-comparison filter in the Sachdeva model variant.

The preservation of complete data integrity at this point allows all future exclusions to be linked to specific physics-based masks instead of random data elimination. Both the experimental data and Simulink outputs together with pump design constraints which included maximum head (H_{max}) and maximum flow rate (q_{max}) values are presented in the appendix for model calibration purposes and future reproducibility needs. The complete dataset is presented in Appendix A to allow full reproduction of the calibration process and enable future researchers to conduct additional sensitivity or validation studies.

Fig. 2 Raw experimental data (inlet pressure [psi], gas fraction, flow rate [gpm], head [ft])

```
%raw data%presure [psig]  alfa    q [gpm] head [ft]

dd=[50  0.1 31  50
50  0.1 37.5    48
50  0.1 44.5    44
50  0.1 53  40
50  0.1 55.5    35
50  0.1 56  31
50  0.1 57.5    27
50  0.1 59  23
50  0.1 61  20
50  0.1 62  16
50  0.1 63  11
50  0.1 65.5    8
50  0.1 71  4
50  0.2 51  16
50  0.2 52  14
50  0.2 54  9
50  0.2 55  7
50  0.2 57  4
50  0.2 61  2
50  0.15    49  24
50  0.15    51  22
50  0.15    52  18
50  0.15    55  13
50  0.15    56  9
50  0.15    58  7
50  0.15    64  3
```

A sample of the raw dataset is shown in Fig. 2, showing data layout and the range of operating conditions used in this work.

2.3 Empirical Model Formulations and Filtering Criteria

The empirical models were implemented through the structures described in their respective literature. The nonlinear equations became linear through natural logarithm operations which allowed OLS regression to estimate coefficients analytically. The following section explains data preparation and transformation procedures for each model together with their explicit log-linear forms.

2.3.1 Turpin [1] Empirical Model

The Turpin model calculates two-phase ESP stage head through an exponential correction of the liquid-only head. To allow parameter identification, the model was linearized using a natural logarithm transformation:

$$\ln\left(\frac{H_m}{H_l}\right) = Ax^2 + \mathrm{B}x \tag{8}$$

where

$$x = \frac{q_g}{p_{in}\, q_l}. \tag{9}$$

The coefficients A and B emerge from ordinary least squares regression analysis of logarithmically processed model. The mathematical relationship between single-phase and two-phase behavior arises through the direct use of the Simulink-derived liquid head (H_l) within the model equation.

All data points that exceed the Turpin surging criteria was filtered out before performing the regression, this is a necessity to achieve the best predictions accuracy, since any points that exceed the surging criteria are considered unstable. The point where ϕ is larger than 1 where all removed, then the linear regression was done on the remaining data points. Cross-validation was not done in this model due to the small number of data points remaining after the filtering process [24].

2.3.2 Sachdeva [10] Pressure-Rise Model (Unfiltered)

The Sachdeva model predicts the pressure increase of each pump stage by applying power-law functions to inlet pressure alongside gas fraction and mixture flow. This formulation was log-transformed to facilitate linear regression:

$$ln(\Delta p_m) = ln(K_2) + E_1\, ln(p_{in}) + E_2\, ln(\alpha) + E_3\, ln(q_l), \tag{10}$$

We then convert the calculated change in pressure to head by the mixture weight density:

$$\gamma_{mix} = \alpha \rho_g + (1 - \alpha)\rho_l, \tag{11}$$

$$H_{pred} = \Delta p \left(\frac{144}{\gamma_{mix}} \right). \tag{12}$$

where 144 is unit conversion factor (psi·ft^2/lbm → ft of head).

The original Sachdeva [10] paper didn't include any surging criteria, so the full dataset was used for the regression, but this introduced a problem that the author worked to solve by suggesting a mask for the data that we will see more about in the next section.

2.3.3 Sachdeva [10] Pressure-Rise Model with Head-Comparison Filter

After many attempts to reduce the MAE and the RMSE of the Sachdeva [10] model, with extensive trial and error and researching the literature, I came to a conclusion that applying a filter that eliminates all the points where the head of mixture is higher than the head of pure liquid obtained from Simulink showed a great improvement in the models accuracy.

The head-comparison mask $H < H_l$ filters out nonphysical results during regression to hone the model on realistic two-phase conditions. Both the filtered and unfiltered versions of the model used the same linearized form of the Sachdeva [10] model, with the only difference being the dataset used.

2.3.4 Zhou and Sachdeva [4] Model

Zhou and Sachdeva [4] use a unified power-law head correlation incorporating known pump design limits, together with a pre-fit mask that removes choked flow data. After taking the logarithm of the original model, we obtained the following linearized equation:

$$\ln\left(\frac{H_m}{H_{max}}\right) = \ln(K) + \alpha E_1 \ln(C_1 p_{in}) + E_2 \ \ln(1-\alpha) + E_3 \ \ln\left(1 - \frac{q_m}{q_{max}}\right). \quad (13)$$

The pump design limits H_{max} and q_{max} can be obtained from the pumps catalog for our certain pump they are 53 ft and 73 gpm, respectively. Prior to regression all the points that don't satisfy the maximum flow criteria introduced in the original literature were removed from the dataset [4]. The model uses data from fully developed flow regimes because of this selection process. Hence we observe that the model achieves improved robustness when working with different gas fractions and pressure values.

2.4 Parameter Identification and Regression Procedure

All parameter estimation was done by using MATLAB's native matrix operations. The log-linear models were written in this general form:

$$\ln(y) = \theta_0 + \theta_1 \ln(x_1) + \theta_2 \ln(x_2) + \ldots, \quad (14)$$

where θ is the vector of coefficients to be identified.

We then used MATLAB's backslash operator to solve the following normal equations:

$$\theta = \left(X^T X\right)^{-1} X^T y. \quad (15)$$

For the construction of regressor matrices for each model, log-transformed variables were used to linearize the relationships. Model coefficients were derived analytically from Eq. (5) and exponentiation. We exponentiated the intercept term where necessary to convert it back to its original scale and maintain the interpretability of the model parameters. This systematic approach ensured robust and efficient parameter estimation while preserving the underlying physical relationships in the data.

2.5 Validation Protocol

The evaluation strategy for this study integrates both data-driven robustness and practical performance insight. The predictive capability of each model is evaluated through appropriate validation techniques to ensure that the fitted parameters generalize well beyond the training dataset.

2.5.1 Performance Metrics (MAE and RMSE)

Two error measurement metrics named mean absolute error (MAE) and root mean squared error (RMSE) serve to evaluate model accuracy.

MAE calculates the mean value of absolute head prediction errors that compare predicted and measured values. This metric provides direct measurements of physical prediction errors in feet since it treats all deviation amounts equally, I take the sum of errors then divide it by the number of points (n),

$$MAE = \frac{1}{n}\sum_{i=1}^{n}\left|H_{pred,\ i} - H_i\right|. \tag{16}$$

The RMSE calculation determines the square root of average squared differences between actual and predicted head values. The penalization of larger errors occurs in this metric because it assigns more weight to extreme deviations compared to MAE,

$$RMSE = \sqrt{\frac{1}{n}\sum_{i=1}^{n}\left|H_{pred,i} - H_i\right|^2}. \tag{17}$$

Each model and configuration receives these metrics for evaluation including both original and filtered data sets whenever applicable. The dual evaluation of models uses both MAE and RMSE to provide a complete analysis of their behavior across different flow conditions and filtering methods.

2.6 Visualization of Results

Several methods were used to visualize the results obtained from the above procedures, visualization help aid the reader with understanding the results. They offer a great method to pass the results of our work to readers simply. Our main focus in this chapter was on performance curves on specific chosen conditions and scatter error residual plots; we will also have plots comparing the MAE and the RMSE of the different models.

2.6.1 Performance Curves

The analysis of experimental ESP behavior under two-phase conditions requires performance curves which compare predicted head values to measured head values. Performance curves which integrate data from all three models of Turpin [1], Sachdeva [10], and Zhou and Sachdeva [4] are created to analyze model behavior under selected operating conditions. The plots display the predicted head values against measured head values while the flow rates change.

The graphs present curves which organize data points according to specific inlet pressure and gas volume fraction (α) combinations to allow comparisons within defined operating conditions. Each curve displays the model's ability to accurately predict two-phase pump performance at that specific condition. The combined display of models on one set of axes reveals variations in fitting quality so users can obtain a complete visual summary in addition to reported error metrics.

2.6.2 Residual Scatter Plots

Head-flow performance plots are accompanied by residual scatter plots to show the prediction error distribution throughout the entire dataset, they are a standard tool for assessing model fit [25]. The difference between predicted and measured head values forms the residual for each data point,

$$Residual_i = H_{pred,i} - H_i. \tag{18}$$

The graphical display shows residuals in relation to the flow rate. Residual plots help users determine if prediction errors exhibit randomness (indicating unbiased model performance) or if there is existing systematic patterns such as higher residuals at low flow rates. The visualization shows residual plots for each model, unfiltered and filtered version for model 2 Sachdeva [10] to evaluate how filtering methods affect error performance.

The visualization framework expands model fidelity assessment through its combination of performance curves and residual analysis techniques.

2.6.3 MAE Comparison Chart

A bar chart displays the Mean Absolute Error (MAE) values to combine and evaluate model performance across all models. A bar chart summarizes MAE values from Turpin [1], Sachdeva [10], and Zhou and Sachdeva [4] models when using literature-based and calculated parameters. A bar representing the filtered Sachdeva model's performance when $H < \text{hl}$ is added to the plot to demonstrate the effect of the author's proposed mask.

The chart displays literature fits and parameter-identified results by distinct color-coded bars. This allows for quick visual comparison of model accuracy improvements achieved through regression. The vertical axis is scaled consistently to ensure visual fairness across models. The visual chart also includes numerical results to help readers understand the general model fitting quality.

2.7 Graphical User Interface (GUI)

The development of a standalone graphical user interface (GUI) through MATLAB App Designer aimed to enhance practical usability of the modeling framework for industrial applications. The GUI enables field engineers and technicians to perform pump head performance estimations through a user-friendly interface which eliminates the need to work with code models or numerical solvers.

The interface presents three input blocks which ask users to enter inlet pressure in psi units and gas volume fraction α between 0 and 1 and flow rate in gallons per minute. The "Estimate Head" button activation triggers the app to calculate and show the predicted head value in feet based on the Zhou and Sachdeva [4] model. The GUI implemented this model because it maintained stable performance after model parameter adjustment and filtering as seen in the results.

The GUI architecture contains all required constants and trained model parameters which enables fast head predictions with minimal user interaction. The actual interface used in this study appears as shown in Fig. 3:

The model integration into a clean and responsive interface provides an accessible link between data-driven modeling and real-world applications in pump diagnostics and design support workflows.

2.8 Summary of Methodological Flow

The chapter started its methodology by importing the full dataset from MATLAB file datt.m. The structure of each model received support from the computation of derived variables including gas and liquid flow rates, mixture density, and normalized pressure and flow terms. The following step involved applying model-specific

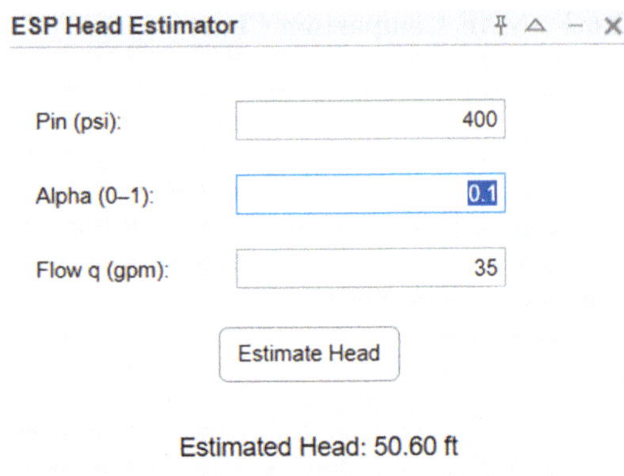

Fig. 3 ESP Head estimator GUI built with MATLAB app designer

filtering criteria to remove unstable and nonphysical data points. The empirical models got log-linear reformulation to achieve closed-form parameter estimation through ordinary least squares regression. The Turpin model received direct fitting because its filtered dataset was too small while 80/20 train/test data splits tested the robustness of other models with adequate data quantity. The evaluation of model performance included mean absolute error and root mean squared error metrics to measure average accuracy and outlier sensitivity together with performance curves and scatter error residual plots. The best-performing model Zhou and Sachdeva [4] was implemented in a graphical user interface that used MATLAB App Designer for development. Engineers can use this tool to predict ESP head under two-phase flow conditions through basic parameter input of pressure, gas fraction and flow rate. The workflow provides a framework for model reproducibility and operational and research application of the calibrated models.

3 Results and Discussion

3.1 Results Overview

This section provides a detailed assessment of the empirical models used for predicting the head performance of electrical submersible pumps (ESPs) when working under two-phase flow conditions. The objective of the results analysis is to verify the precision and reliability of each model after parameter identification while checking the applicability of the proposed methods across different flow patterns and different inlet pressures and gas volumes. The section assesses the effectiveness of the data filtering method presented by the author and its impact on the performance of the Sachdeva [10] correlation.

The analysis focuses on three essential performance indicators which include Mean Absolute Error (MAE), Root Mean Squared Error (RMSE), and residual patterns. These are displayed through a combination of summary bar charts together with predicted-versus-measured comparison plots and residual scatter plots. They provide both numerical and descriptive insights about model performance after parameter optimization.

The analysis of performance trends uses numerical data as well as theoretical pump physical principles. The impact of increasing gas content together with rising flow rates on head degradation is analyzed through the lens of fluid mechanics principles. The academic objective of showing comprehension exceeds statistical performance results.

The filtered Sachdeva model variant receives particular attention since it implements a new masking criterion which the author suggests to eliminate data points where measured two-phase head exceeds the single-phase baseline. The contribution stems from detailed dataset assessment combined with domain knowledge which leads to a practical improvement of prediction accuracy and visual coherence.

Performance curve overlays present the results at 400 psi with $\alpha = 0.4$ and 100 psi with $\alpha = 0.15$. These visualizations enable users to evaluate experimental head-flow trends directly against all models within specific operating conditions. The models demonstrate their best generalization capabilities for real ESP behavior through this method.

The section presents the fitted parameters in a table format before explaining the rationale behind choosing the Zhou & Sachdeva model for the implemented GUI application. The GUI uses this model because it offers a good balance of generality and accuracy and simplicity in implementation which results validate.

The result section provides a technical evaluation of modeling approaches that combines data evidence with physical significance to demonstrate the author's involvement in enhancing and deploying the methodology.

3.2 *Mean Absolute Error (MAE) Analysis*

After parameter identification, the mean absolute error (MAE) functions as the primary assessment criterion for evaluating model prediction accuracy. MAE gives an understandable measure of the average prediction difference in feet which is essential for operational settings like pump selection and diagnostics.

MAE results from the four evaluated models indicate performance patterns across Turpin Sachdeva (unfiltered) Sachdeva (filtered with $H < H_l$ mask) and Zhou and Sachdeva. The bar chart in Fig. 4 demonstrates that the Sachdeva model without filtering produces the largest prediction errors. The inconsistent physical data points cause negative effects on the regression accuracy.

The implementation of the head-comparison mask ($H < H_l$) proposed by the author results in a significant improvement of MAE performance. The filtered Sachdeva model shows substantially reduced average errors than its unfiltered version which supports the theory that removing nonphysical points leads to a better

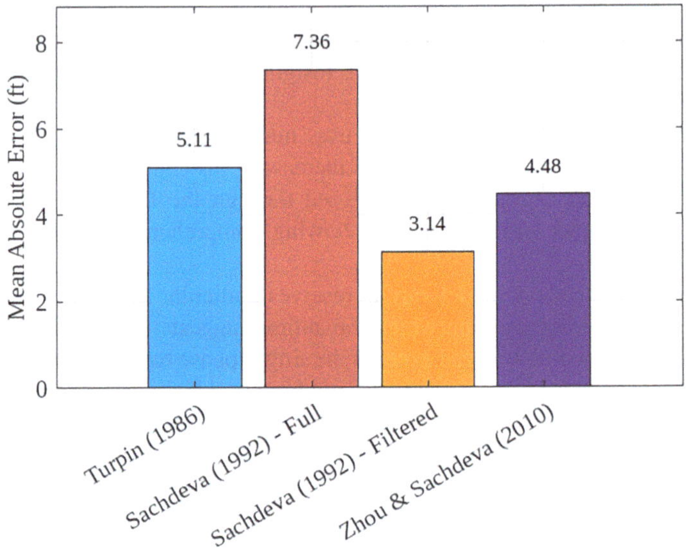

Fig. 4 MAE comparison bar chart

fit. The MAE reduction proves the value of the author's mask as well as the usefulness of purposeful preprocessing techniques for empirical modeling.

The Turpin model performs reasonably well considering its basic structure although it achieves moderate MAE values because it was trained on a limited filtered dataset. The model functions as a reference point for evaluating performance stability in flow regimes.

The Zhou and Sachdeva model demonstrates the lowest MAE value which proves its ability to function well under different operating scenarios. The model demonstrates outstanding performance because it combines advanced structural elements with the advantages of being trained on filtered data. The Zhou model achieves this strong outcome which supports its selection for developing the practical GUI tool.

The research demonstrates that parameter identification increase prediction accuracy when used with physically relevant data filtering methods. The MAE metric enables researchers to evaluate model reliability in realistic two-phase conditions through straightforward interpretation.

3.3 Root Mean Squared Error (RMSE) Analysis

RMSE represents a helpful alternative to standard metrics by specifically showing the size of major prediction mistakes. The squaring of each residual in RMSE before averaging produces results which reveal model sensitivity to outliers and provides evidence about the reliability and robustness of each empirical fit.

The unfiltered Sachdeva model shows the highest RMSE value compared to all models studied in this analysis. The high prediction errors stem from model exposure to physically wrong data points which cause major discrepancies between predicted head and two-phase measurement values. The large prediction errors show that applying the model without consideration results in unstable predictions which reduce its predictive capability.

The Sachdeva model version with the head-comparison mask ($H < \mathrm{hl}$) proposed by the author leads to a better RMSE result than the unfiltered version. The RMSE reduction proves that the proposed filtering method successfully removes nonphysical conditions which otherwise would distort the regression model. The better RMSE results show that the proposed filter improves both overall prediction alignment and minimizes severe mismatches.

The Turpin model maintains a good RMSE value because of its basic structure along with its training on a reduced set of filtered data. The model achieves stable predictions because it uses the liquid-only head as its reference point which prevents unpredictable results. The model achieves stable performance even with its basic design structure.

The Zhou and Sachdeva model produces one of the lowest RMSE values which proves its capability to maintain robustness in multiple input scenarios. The model's hybrid structure and normalized variable approach make it resistant to large residuals. The results support its deployment in practical applications thus validating the MATLAB GUI development within this chapter.

Models that receive proper calibration through filtering and informed formulation achieve better stability in their predictions. RMSE functions as a diagnostic tool to identify models which both achieve good average fit performance and handle outliers effectively (Fig. 5).

3.4 *Predicted vs. Measured Head Trends*

The report contains a direct comparison of predicted two-phase head values and measured values in all models. The diagrams show how well each model replicates the real experimental results.

The unfiltered Sachdeva model displays the largest amount of data points that deviate from the diagonal reference line because it shows systematic overestimation together with inconsistencies. The model shows its difficulty in handling unrealistic or physically contradictory values through its visual results that match its high error metrics (Fig. 6).

The filtered Sachdeva model shows reduced scatter while its points become more aligned with the 1:1 line when the author implements the head-based filter ($H < \mathrm{hl}$). The filtering method enhances both structural predictability of the model and error metrics in the input domain according to statistical results (Fig. 7).

The Turpin model maintains a compact data point distribution that closely follows the reference line because it operates within stable low-gas conditions

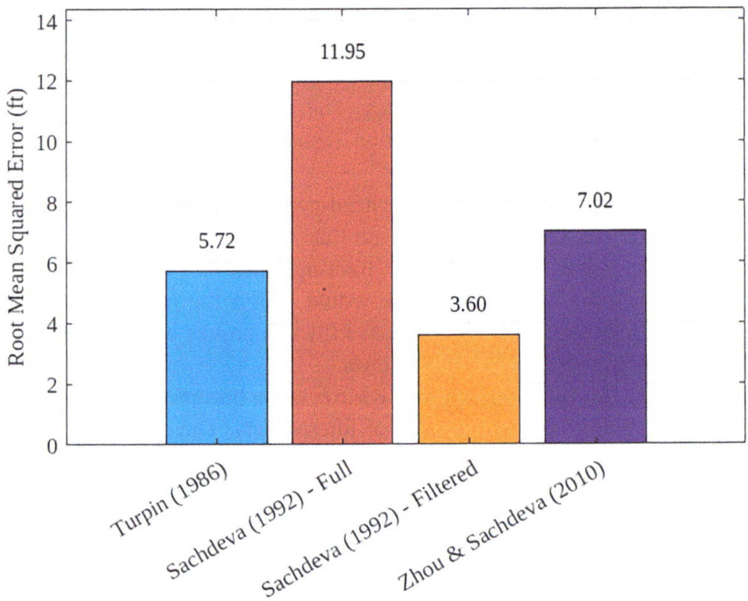

Fig. 5 RMSE comparison bar chart

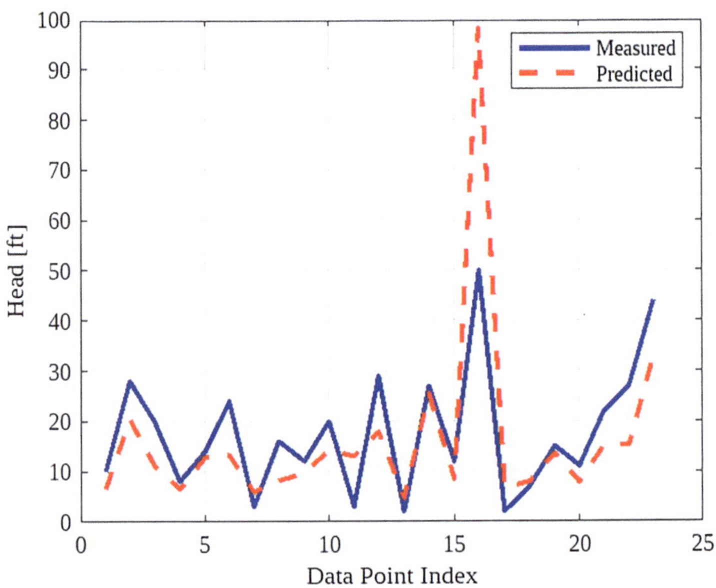

Fig. 6 Predicted head VS experimental head for Sachdeva [10] unfiltered

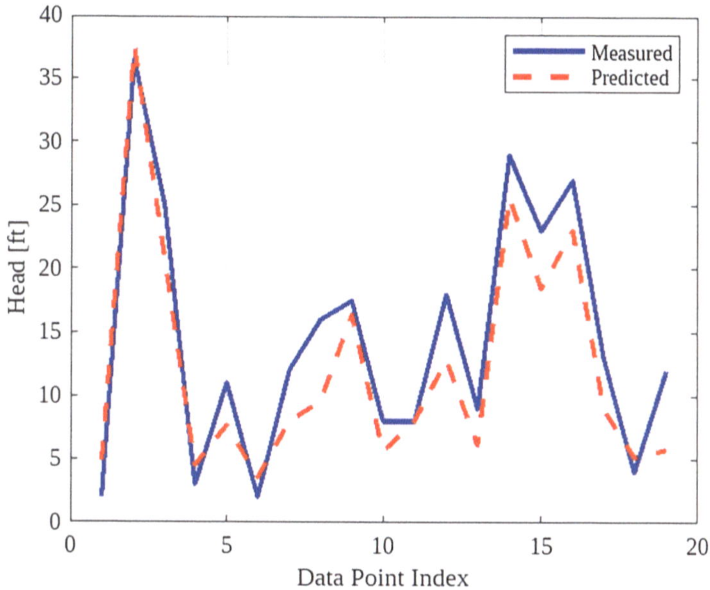

Fig. 7 Predicted head VS experimental head for Sachdeva [10] with author-suggested filter

based on surge criteria. The filtered data subset shows a consistent pattern from the Turpin model except for some variability in the high flow rate region because the exponential form becomes less sensitive to gas effects (Fig. 8).

The Zhou and Sachdeva model shows the most uniform distribution of predicted values which remain tightly grouped throughout the entire data range. The model uses normalized flow and pressure terms to achieve generalized performance across various gas fractions and inlet pressures. The model demonstrates strong practical application potential because visual consistency is essential for building trust in engineering models (Fig. 9).

The visual analysis of predicted vs. measured head values enhances the numerical error metrics by showing both the magnitude of model errors and their ability to replicate observed ESP behavior patterns.

3.5 Residual Analysis

The analysis of residuals serves as an alternative approach to evaluate model performance by showing how prediction errors change in relation to operating conditions including flow rate, gas volume fraction, and inlet pressure. The following section shows residual scatter plots for each of the four fitted models to help both detect patterns and identify patterns of systematic bias and error clusters.

The residual plot of the Turpin model display residuals which maintain a thin distribution near zero without any detectable systematic bias. The model maintains

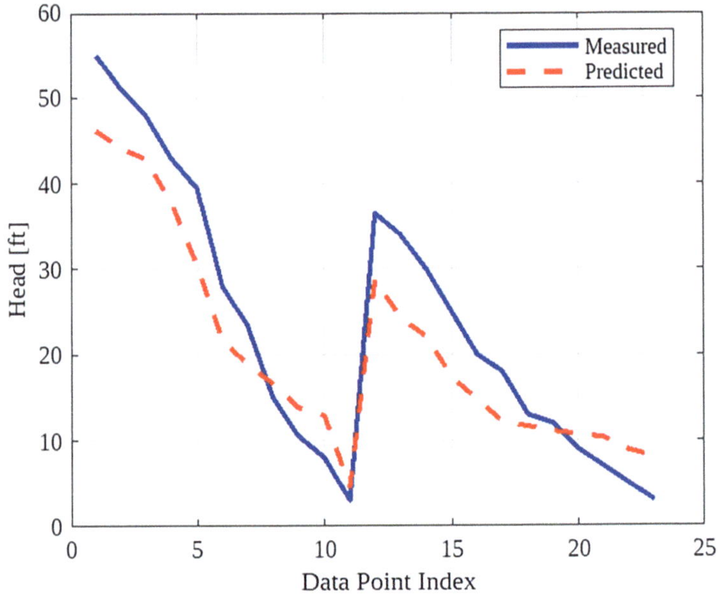

Fig. 8 Predicted head VS experimental head for Turpin [1]

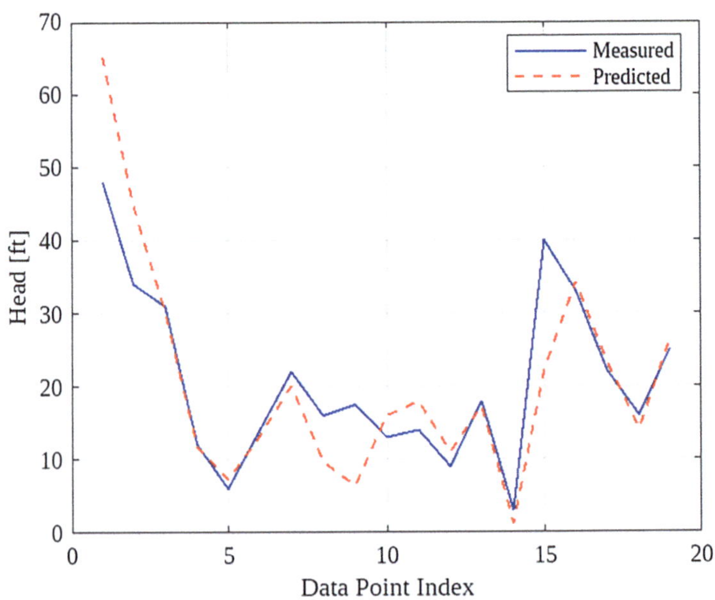

Fig. 9 Predicted head VS experimental head for Zhou and Sachdeva [4]

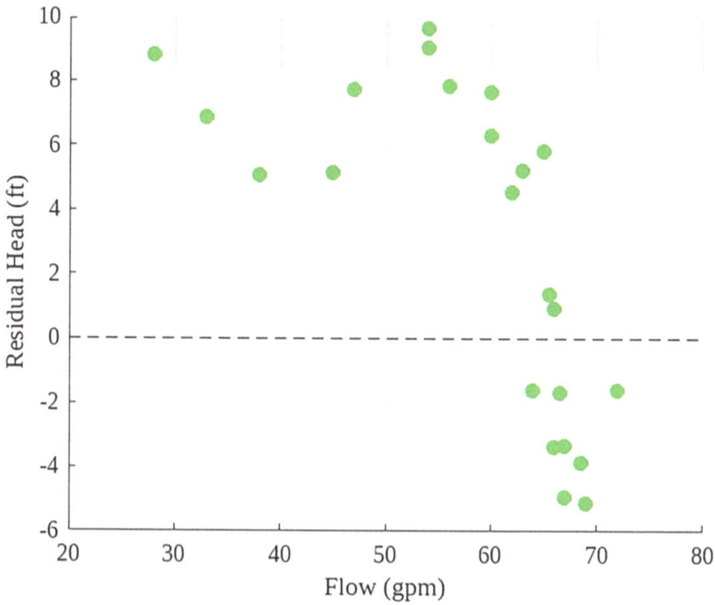

Fig. 10 Residuals Plot: Turpin [1]

consistent performance throughout the operational envelope because the $\varphi < 1$ surge criterion maintains flow stability. Underprediction occurs at high flow rates because gas interference exceeds the model's design boundaries (Fig. 10).

The unfiltered Sachdeva model produces the most dispersed residual distribution which shows both positive and negative bias areas across different flow rate intervals. Systematic patterns in the residuals confirm that this version fails to achieve necessary robustness when processing the dataset with physically implausible measurements (Fig. 11).

The filtered Sachdeva model ($H <$ hl mask) shows significant improvement in residual distribution. The data points distribute evenly around zero and most extreme values disappear from the scatter plot. The filtering approach confirmed by the author reduces both average prediction errors and improves the regression model's structural quality across different operating points (Fig. 12).

The residual pattern of the Zhou and Sachdeva model cluster tight around zero with minimal signs of flow-dependent distortion or bias. The model achieves successful performance throughout the entire dataset because it uses effective normalization techniques and mathematical structure which adapt well to the nonlinear behaviors of flow and pressure conditions. Random patterns in its residuals make this model suitable for GUI implementation.

Residual plots demonstrate that intelligent filtering techniques together with robust functional forms make substantial contributions to model reliability. The observed trends demonstrate the critical need for residual-based diagnostics in choosing models for predictive engineering applications (Fig. 13).

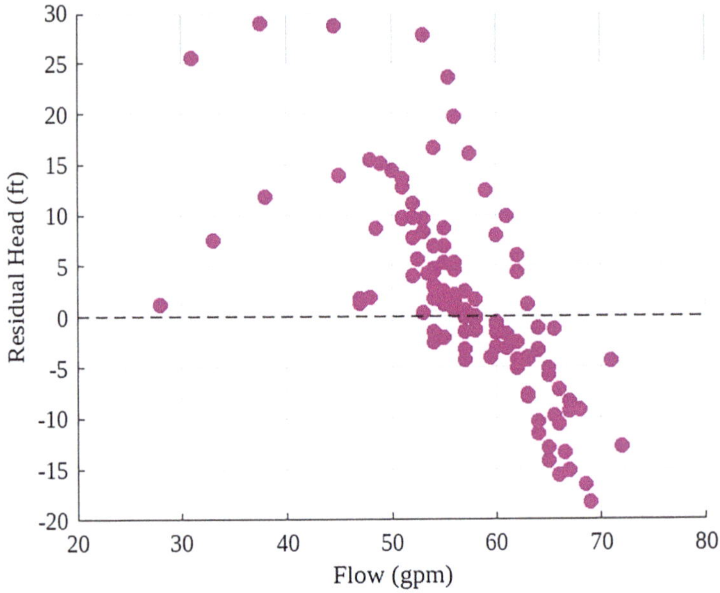

Fig. 11 Residuals Plot: Sachdeva [10] unfiltered

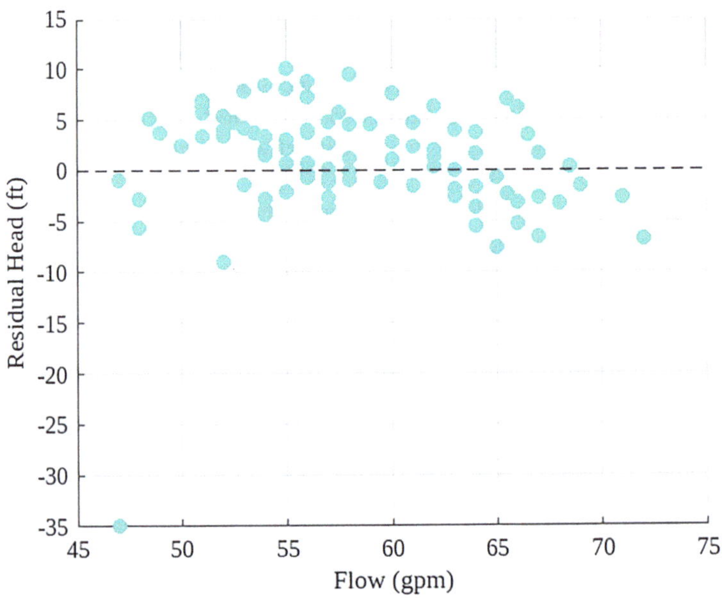

Fig. 12 Residuals Plot: Sachdeva [10] with author-suggested filter

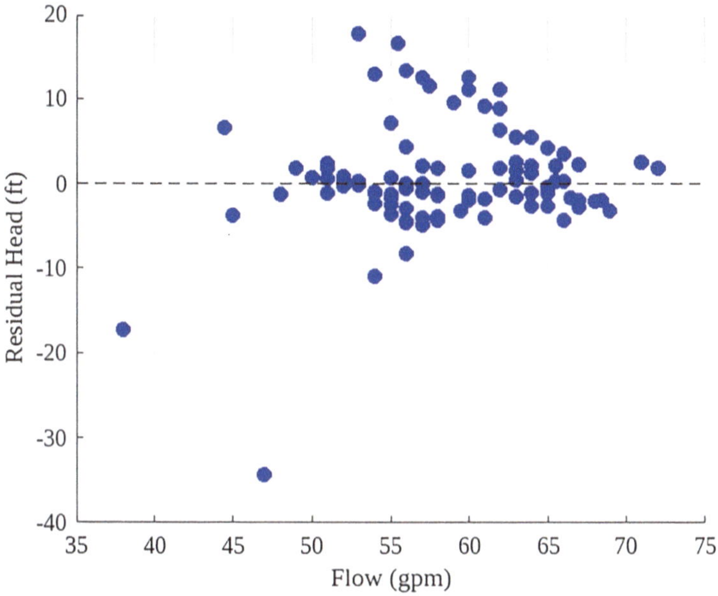

Fig. 13 Residuals Plot: Zhou and Sachdeva [4]

3.6 Performance Curve Comparisons

The evaluation of physical realism and operational relevance of fitted models becomes easier through the use of performance curve comparisons together with error-based metrics and residual trends. The analysis includes head-versus-flow rate curves of all models for two operating conditions: (1) High inlet pressure with moderate gas fraction at 400 psi and $\alpha = 0.4$ and (2) Low inlet pressure with low-gas fraction at 100 psi and $\alpha = 0.15$.

When operating at 400 psi and $\alpha = 0.4$, the observed head-flow trend shows a clear decrease in head as the flow rate rises which indicates typical gas interference effects on ESP performance. The unfiltered Sachdeva model shows significant departure from the measured trend during this regime since it does not consider physical head limits despite its sensitivity to gas volume. The filtered Sachdeva model shows better agreement with experimental data through its ability to track the curvature of measured points because the head-based masking technique eliminates physically unrealistic behaviors (Fig. 14).

The Zhou and Sachdeva model demonstrates the most precise fit to experimental data points during both ascending and descending segments by modeling the nonlinear degradation that occurs with increased flow and gas entrainment. The model demonstrates consistent results throughout different flow ranges which validates its engineering application and makes it a suitable candidate for GUI integration.

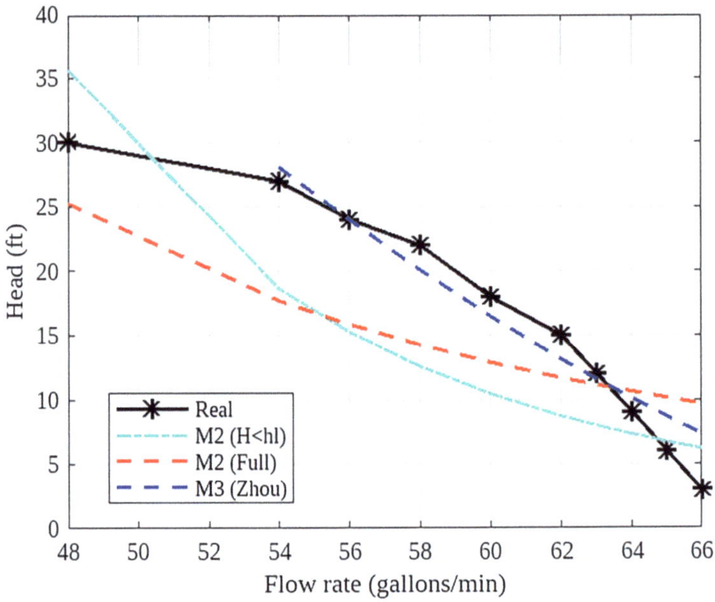

Fig. 14 Performance curves at 400 psi, $\alpha = 0.4$

The models show better agreement with the head-flow relationship at 100 psi and $\alpha = 0.15$ but distinct variations persist. The unfiltered Sachdeva model shows significant differences at middle-range flow rates while both the filtered Sachdeva variant and Turpin model demonstrate better curve adherence. The Zhou and Sachdeva model demonstrates outstanding precision throughout the entire curve including the high flow regions where degradation becomes more severe (Fig. 15).

The performance curves serve as visual proof that each model demonstrates strong generalization capabilities especially when operating under difficult two-phase conditions. These models show both their precision in predictions and their capability to duplicate actual ESP behaviors in actual field operations. The Turpin model did not receive comparison because the $\varphi < 1$ filtering removed most data points which prevented it from producing a valid curve throughout the operating range.

3.7 Summary of Fitted Parameters and GUI Model Justification

The last section unites the essential model coefficients obtained from regression with their practical implications for selecting the GUI tool's model.

The fitted parameters of each model appear in Table 4 which includes the Sachdeva and Zhou power-law exponents (E1, E2, E3) together with the base

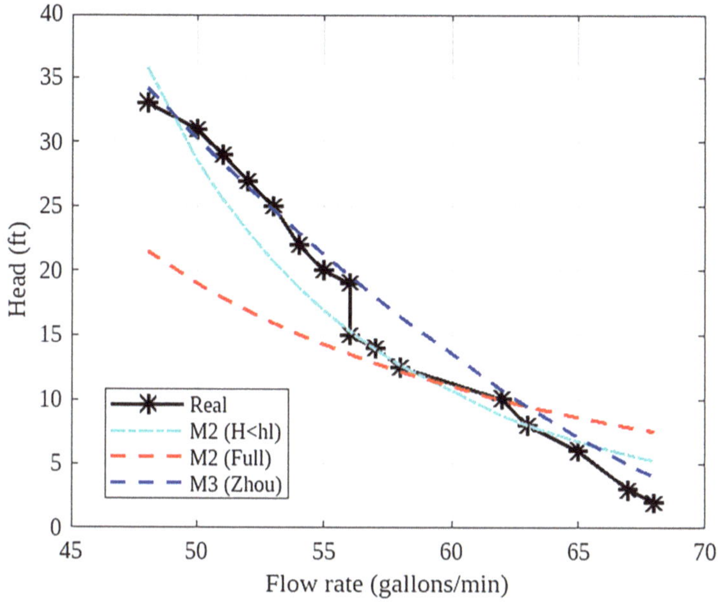

Fig. 15 Performance curves at 100 psi, $\alpha = 0.15$)

Table 4 New identified parameters by OLS

Model	K	E1	E2	E3
Turpin [1]	Na	5337800.0	−6002.7	Na
Sachdeva [10]	67.2762	0.6131	−2.0972	−3.0203
Sachdeva [10] - Filtered	108650	0.6602	−3.2140	−5.5046
Zhou and Sachdeva [4]	4.3271	3.5799	16.9516	1.5864

coefficient (K). The Turpin model contains A and B coefficients instead of K and exponents since it employs a quadratic exponential form. The parameters were estimated through ordinary least squares after data transformation and filtering for a uniform estimation method among models.

The model parameters show structural divergence between models while showing their reaction patterns to changing input variables. The filtered Sachdeva model obtains high K value because of removing outlier data points that compressed its regression surface. The model becomes more effective at displaying natural physical relationships after removing extreme data points because it produces a steeper curve. The observed behavior follows standard practices in empirical modeling since data refinement enhances the sensitivity of model parameters toward fundamental physical effects. The Zhou model's parameters demonstrate an optimal combination of responses to inlet pressure and gas fraction and flow rate which matches its consistent results in both residual and visual tests.

The comprehensive evaluation in this section led to the selection of the Zhou and Sachdeva [4] model for the MATLAB GUI use. The model showed consistent behavior across all data points and needed minimal parameter tuning to achieve low errors while giving consistent visual results under different operating conditions.

The model selection represents a balance between model complexity and reliability because it supports flexible nonlinear modeling of pressure and gas and flow interactions while maintaining interpretability and computational efficiency. Its features make it appropriate for deployment in field-facing tools that require accurate fast results and user trust.

4 Conclusion

4.1 Conclusion

This research conducted a detailed evaluation and calibration process for three fundamental empirical models that predict ESP head behavior under gas–liquid two-phase petroleum flow conditions. The research investigated these conditions because petroleum wells with multiphase flow present operational challenges when free gas reduces ESP performance thus requiring accurate head estimation for design and optimization and troubleshooting purposes.

Turpin [1], Sachdeva [10], and Zhou and Sachdeva [4] represent the three selected models due to their high visibility in literature alongside their different structural approaches. All models received MATLAB implementation which underwent standardized calibration and validation testing. The modeling structure incorporated features for transparent modeling and reproducible results and physical constraints maintenance together with computational efficiency suitable for field application.

The main contribution of this research involved implementing data filtering which matched the requirements of each selected model. The Turpin model included a surge criterion filter ($\varphi < 1$) to select stable flow conditions while the Sachdeva model underwent two different analyses: its standard form and an evaluation using the new head-comparison mask ($H <$ hl) created by the author. The Zhou and Sachdeva model utilized a critical flow ratio mask from existing literature. The purpose of these specific filters was to direct the calibration process toward meaningful data points while blocking the introduction of instability-related and choked flow and measurement inconsistencies.

To achieve linearizable forms the models needed reformulation through logarithmic transformations before the OLS method estimated their coefficients. This regression method produced solutions that could be represented through closed-form equations which improved both methodological understanding and implementation simplicity. All evaluation procedures relied on actual data, while literature-based parameter values were used only for benchmarking purposes.

The evaluation of model accuracy used two error metrics which consisted of Mean Absolute Error (MAE) and Root Mean Squared Error (RMSE). The results confirmed that the identification of model parameters resulted in better accuracy than when using literature-based coefficients. The MAE and RMSE results demonstrated that unfiltered models particularly the original Sachdeva formulation demonstrated systematic deviations and overfitting which led to unrealistic head predictions in gas-rich conditions.

The head-comparison mask developed in this research proved highly successful. The mask function eliminated measurement conditions that biased regression outcomes by removing data points when two-phase head measurements exceeded single-phase liquid head calculations from Simulink. The filtered Sachdeva model achieved better MAE and RMSE results and developed symmetric residuals which spread evenly around zero thus producing a more stable fit.

Among all models, the Zhou and Sachdeva correlation emerged as the most robust and reliable. The model achieved the lowest prediction errors throughout all evaluation metrics and generated residual plots with minimal bias. The structural implementation of normalized flow and pressure terms in the model enabled it to generalize well across different operational settings. The performance curves under representative conditions (400 psi, $\alpha = 0.4$ and 100 psi, $\alpha = 0.15$) showed that the model effectively duplicated physical head reduction patterns as flow rates increased.

The Turpin model remained conceptually appealing due to its simple implementation yet suffered from insufficient data points that fulfilled its surge criterion thus limiting its application. The model received exclusion from performance curve visualizations because it lacked enough data points to create meaningful trends. The model showed steady residual patterns throughout stable operating areas which supports its use as an effective specialized tool for particular low-gas conditions.

The research includes the creation of a MATLAB graphical user interface (GUI) for the Zhou and Sachdeva model. Users can use the MATLAB App Designer GUI to enter inlet pressure and gas fraction and flow rate values which generates immediate head estimates for the ESP. This tool offers technicians and field engineers a practical interface to work with by simplifying the complex model into an easy-to-use platform. The inclusion of this tool in the chapter shows a dedication to academic excellence together with practical applications in the real world.

This chapter provides both theoretical and practical insights about applying empirical ESP head models in petroleum flow systems with two phases. The source code maintained its MATLAB Live Script format which enables researchers to easily enhance model functionality and validate or reuse the code. The research implements methodological breakthroughs alongside detailed parameter selection procedures and field-oriented tool development to create an applicable framework that enhances ESP performance prediction.

4.2 *Recommendations for Future Work*

This chapter shows strong results through data evidence, but multiple future research directions exist for development purposes:

1. **Expanded Dataset Collection**

The experimental dataset used for the current research includes CO_2/diesel mixtures only. Future research should use a wide array of experimental datasets that include various gas-liquid mixtures and pressure conditions and pump types. The use of diverse data would enhance model generalization together with improved statistical validation results.

2. **Integration with Mechanistic Models**

Future studies should combine empirical models with mechanistic models for developing hybrid approaches that preserve speed and simplicity from empirical models yet maintain physical accuracy of mechanistic models. The inclusion of gas slip mechanisms alongside bubble coalescence and flow pattern changes would improve model realism without causing substantial increases in computational requirements.

3. **Advanced Regression Techniques**

The research adopted ordinary least squares because it provided clear results but future investigations should test weighted least squares and robust regression and Bayesian estimation methods. The techniques provide better performance when analyzing datasets containing heteroscedasticity and non-normal error distributions.

4. **Sensitivity and Uncertainty Analysis**

The research focused on achieving the best possible fit with visual representations while minimizing prediction errors. Future studies need to perform sensitivity tests which identify the variables that influence model output predictions the most. Enhanced decision-making in safety-critical applications will benefit from uncertainty prediction methods.

5. **Incorporating Real-Time Field Data**

The present models depend on offline datasets for calibration purposes. The upcoming research should use real-time sensor data from operating wells to validate and update the models. The combination of model adaptation with continuous performance tracking through real-time data would lead to substantial field utility improvements.

6. **Field Deployment and User Testing of GUI**

The field deployment of the GUI represents a vital next step although the interface was developed and tested in MATLAB. The design process needs to receive feedback from technicians and engineers through interface testing before designers can create the final version. The investigation should assess the possibility of integrating the system with web-based tools or mobile applications.

7. **Cross-Model Comparative Frameworks**

The research analysis only examined three different models. A larger comparative analysis including empirical models and machine learning models would determine the position of each model in relation to others in performance space. Research findings will establish optimal practices for various operational environments through comparative assessments.

8. **Publication and Open-Access Toolkits**

The research needs to reach maximum impact so future studies must publish the GUI along with model code in open-access repositories. Open-source availability of the research will boost transparency levels while promoting adoption and permitting multiple researchers and practitioners to work together for improvement.

4.3 Final Remarks

This research provides essential and accurate tools to estimate ESP head under two-phase flow conditions that fulfill current petroleum engineering needs. The combination of experimental data analysis with regression application and a novel filtering method resulted in a modeling framework that delivers useful operational benefits.

The decision to choose the Zhou and Sachdeva model was based on both statistical performance and operational considerations as well as the model's robustness. The MATLAB-based GUI development enables the transition between theoretical modeling and field application to benefit engineers and technicians and decision-makers for ESP system optimization.

This research delivers present solutions as well as a starting point for future advancement and development. The models and tools established in this study can develop into essential elements of upcoming petroleum production optimization workflows through continuous research and testing efforts.

Appendices

MATLAB live script screenshots:

Parameter Identification of Empirical Models for ESP Head Estimation.

1.1. Data Import

```
set(0, 'DefaultAxesFontName', 'Times New Roman');
set(0, 'DefaultTextFontName', 'Times New Roman');

run ('datt.m');    % extract dataset from M-file
pin = dd(:,1);  % extract inlet pressure into pin
alpha = dd(:,2);  % extract gas void fraction into alpha
q = dd(:,3);  % extract flow rate into q
H = dd(:,4);  % extract head into H
```

1.2. Extract hl from Simulink and plot to compare

```
qi = q;
sim('cur');
hl = hl(:);

plot(H, 'b-', 'LineWidth', 2);       % H in blue line
hold on
plot(hl, 'r--', 'LineWidth', 2);      % hl in red dashed line
xlabel('Data Point Index');
ylabel('Head [ft]');
legend('Head of Mixture (Hm)', 'Head of Liquid (hl)');
```

2. Model 1 (Turpin, 1986) Fitting

2.1. Compute phase flow rates from mixture data

```
ql = (1 - alpha) .* q;  % extract liquid flow into ql
qg = alpha .* q;        % extract gas flow into qg, the dot multiplies two lists one by one

```

2.2. filter stable data for model 1(Turpin, 1986)

```

phi1  = 2000*(qg./ql)./(3*pin);    % compute surge indicator for Model 1
idx1  = phi1 < 1;                  % logical mask of stable points
pin1  = pin(idx1);                 % extract stable inlet pressures
ql_M1   = ql(idx1);                  % extract stable liquid flows
qg_M1   = qg(idx1);                  % extract stable gas flows
H_M1    = H(idx1);                   % extract stable mixture heads
hl_M1   = hl(idx1);                  % extract stable liquid heads
```

3.1 calculating estimated mixture density for presuure to head conversion

```
% 1) ΔP from Model 2 (Sachdeva 1992)
K    = 1.154562;
E1   = 0.943308;
E2   = -1.175596;
E3   = -1.300093;
dp = K .* pin.^E1 .* alpha.^E2 .* q.^E3;   % [psi per stage]
atm    = 14.7;              % psi


% 2) Total absolute pressure seen by the gas [psia]

P_tot = pin + dp + atm;                  % [psia = psig + 14.7]

% 3) Gas density d_g [lb/US gal]
R       = 1130;                          % ft·lbf/(slug·°R)
T       = 620;                           % °R
conv    = 32.17405/7.48052;              % slug/ft³ → lb/US gal
d_g     = (P_tot .* 144) ./ (R .* T) .* conv;

% 4) Liquid density d_l [lb/US gal] (given from density tables)
d_l     = 6.943;

% 5) Mixture density d_mix [lb/US gal]
d_mix  = alpha .* d_g + (1 - alpha) .* d_l;
```

Dataset:

presure [psig]	alfa	q [gpm]	head [ft]
50.0	0.1	31.0	50.0
50.0	0.1	37.5	48.0
50.0	0.1	44.5	44.0
50.0	0.1	53.0	40.0
50.0	0.1	55.5	35.0
50.0	0.1	56.0	31.0
50.0	0.1	57.5	27.0
50.0	0.1	59.0	23.0
50.0	0.1	61.0	20.0
50.0	0.1	62.0	16.0
50.0	0.1	63.0	11.0
50.0	0.1	65.5	8.0
50.0	0.1	71.0	4.0
50.0	0.2	51.0	16.0
50.0	0.2	52.0	14.0
50.0	0.2	54.0	9.0
50.0	0.2	55.0	7.0
50.0	0.2	57.0	4.0
50.0	0.2	61.0	2.0
50.0	0.15	49.0	24.0
50.0	0.15	51.0	22.0
50.0	0.15	52.0	18.0
50.0	0.15	55.0	13.0
50.0	0.15	56.0	9.0
50.0	0.15	58.0	7.0
50.0	0.15	64.0	3.0
50.0	0.3	55.0	2.0
50.0	0.3	54.0	6.0
50.0	0.3	53.5	8.5
50.0	0.3	52.5	10.0
50.0	0.4	54.0	1.0
50.0	0.4	54.0	2.0
50.0	0.4	53.0	4.0
100.0	0.1	28.0	55.0
100.0	0.1	33.0	51.0
100.0	0.1	38.0	48.0
100.0	0.1	45.0	43.0
100.0	0.1	54.0	39.5
100.0	0.1	60.0	28.0
100.0	0.1	62.0	23.5
100.0	0.1	64.0	15.0
100.0	0.1	66.0	10.5

(continued)

presure [psig]	alfa	q [gpm]	head [ft]
100.0	0.1	67.0	8.0
100.0	0.1	72.0	3.0
100.0	0.15	48.0	33.0
100.0	0.15	50.0	31.0
100.0	0.15	51.0	29.0
100.0	0.15	52.0	27.0
100.0	0.15	53.0	25.0
100.0	0.15	54.0	22.0
100.0	0.15	55.0	20.0
100.0	0.15	56.0	19.0
100.0	0.15	56.0	15.0
100.0	0.15	57.0	14.0
100.0	0.15	58.0	12.5
100.0	0.15	62.0	10.0
100.0	0.15	63.0	8.0
100.0	0.15	65.0	6.0
100.0	0.15	67.0	3.0
100.0	0.15	68.0	2.0
100.0	0.2	51.0	22.0
100.0	0.2	53.0	20.0
100.0	0.2	55.0	18.0
100.0	0.2	56.0	13.0
100.0	0.2	58.0	12.0
100.0	0.2	60.0	9.0
100.0	0.2	61.0	8.0
100.0	0.2	64.0	8.0
100.0	0.3	48.5	18.0
100.0	0.3	55.0	16.5
100.0	0.3	56.0	13.0
100.0	0.3	57.0	10.0
100.0	0.3	57.0	8.0
100.0	0.3	58.0	6.0
100.0	0.3	59.5	3.0
100.0	0.4	54.0	11.0
100.0	0.4	55.0	9.0
100.0	0.4	56.0	7.0
100.0	0.4	57.0	3.0
100.0	0.4	57.0	2.0
400.0	0.3	47.0	36.5
400.0	0.3	54.0	34.0
400.0	0.3	56.0	30.0
400.0	0.3	60.0	25.0
400.0	0.3	63.0	20.0

(continued)

presure [psig]	alfa	q [gpm]	head [ft]
400.0	0.3	65.0	18.0
400.0	0.3	65.5	13.0
400.0	0.3	66.0	12.0
400.0	0.3	66.5	9.0
400.0	0.3	67.0	7.0
400.0	0.3	68.5	5.0
400.0	0.3	69.0	3.0
400.0	0.4	48.0	30.0
400.0	0.4	54.0	27.0
400.0	0.4	56.0	24.0
400.0	0.4	58.0	22.0
400.0	0.4	60.0	18.0
400.0	0.4	62.0	15.0
400.0	0.4	63.0	12.0
400.0	0.4	64.0	9.0
400.0	0.4	65.0	6.0
400.0	0.4	66.0	3.0
400.0	0.5	47.0	28.0
400.0	0.5	52.0	27.0
400.0	0.5	54.0	25.0
400.0	0.5	57.0	21.0
400.0	0.5	60.0	17.5
400.0	0.5	62.0	14.0
400.0	0.5	63.0	10.0
400.0	0.5	64.0	6.0
400.0	0.5	65.0	3.0

Simulink model:

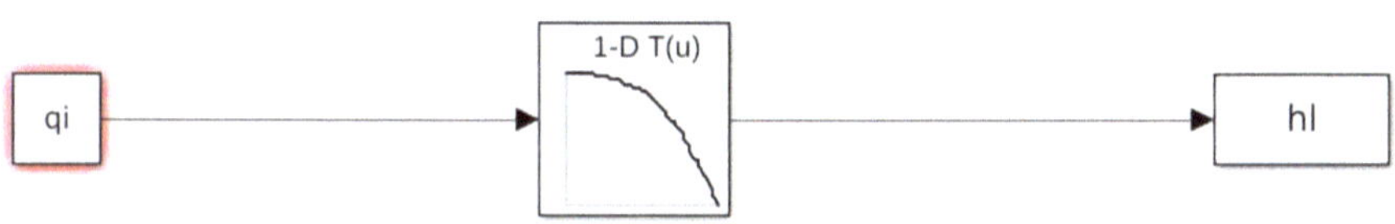

References

1. Turpin J, Lea J, Bearden J (1986) Gas-liquid flow through centrifugal pumps-correlation of data. https://oaktrust.library.tamu.edu/items/10910d3b-b55f-4d91-b50f-8cfd85303722
2. Weriono W, Junaidi A, Indarto K (2024) Effect of venting equipment installation on oil production stability. Isomase.Org, 68. https://isomase.org/Journals/index.php/jomase/article/view/377

3. Nguyen T (2020) Artificial lift methods design, practices, and applications. https://doi.org/10.1007/978-3-030-40720-9_3
4. Zhou D, Sachdeva R (2010) Simple model of electric submersible pump in gassy well. J Pet Sci Eng 70(3–4):204–213. https://doi.org/10.1016/J.PETROL.2009.11.012
5. Mohammadzaheri M, AlQallaf A, Ghodsi M, Ziaiefar H (2018) Development of a Fuzzy Model to estimate the head of gaseous petroleum fluids driven by electrical submersible pumps. Fuzzy Inf Eng 10(1):99–106. https://doi.org/10.1080/16168658.2018.1509523
6. Mohammadzaheri M, Tafreshi R, Khan Z, Ghodsi M, Franchek M, Grigoriadis K (2020) Modelling of petroleum multiphase flow in electrical submersible pumps with shallow artificial neural networks. Ships Offshore Struct 15(2):174–183. https://doi.org/10.1080/17445302.2019.1605959
7. Ali A, Si Q, Wang B, Yuan J, Wang P, Rasool G, Shokrian A, Ali A, Zaman MA (n.d.) Comparison of empirical models using experimental results of electrical submersible pump under two-phase flow: Numerical and empirical model validation. Iopscience.Iop.Org. https://doi.org/10.1088/1402-4896/AC6E96/META
8. Mohammadzaheri M, Tafreshi R, Khan Z, Ziaiefar H, Ghodsi M, Franchek M, Grigoriadis K (2019) Modelling of electrical submersible pumps for petroleum multiphase fluids, an intelligent approach supported by a critical review and experimental results. J Eng Res 16(2):77–86. https://doi.org/10.24200/TJER.VOL16ISS2PP77-86
9. Ali A, Yuan J, Deng F, Wang B, Liu L, Si Q, Buttar NA (2021) Research progress and prospects of multi-stage centrifugal pump capability for handling gas–liquid multiphase flow: comparison and empirical model validation. Energies 14(4):0896. https://doi.org/10.3390/EN14040896
10. Sachdeva R, Doty DR, Schmidt Z (1992) Performance of axial electric submersible pumps in a gassy well: 201–212. https://doi.org/10.2118/24328-MS
11. Zhu H, Yu H, Sun Q, Wang Q, Jing H, Abdikadyrov R (2025) Deep learning-based performance prediction of electric submersible pumps under viscous and gas–liquid flow conditions. Machines 13(2):0135. https://doi.org/10.3390/MACHINES13020135
12. Lea JF, Bearden JL (1982) Effect of gaseous fluids on submersible pump performance. JPT, J Pet Technol 34(12):2922–2930. https://doi.org/10.2118/9218-PA
13. Ortiz Pinilla, J, & Gil, D (2014). Transformaciones logarítmicas en regresión simple. Comunicaciones en Estadística, 7(1), 80–98. https://doi.org/10.15332/s2027-3355.2014.0001.06
14. Morrison G, Carvajal N, Saleh R, Bai C (2015) The measured impact of erosion on the rotodynamic and performance characteristics of a mixed flow ESP. https://doi.org/10.21423/R17W6P
15. AL-Hejjaj MA, Sadeq DJ, Al-Fatlawi O (2023) A review of the electrical submersible pump development chronology. Iraqi J Chem Pet Eng 24(2):123–135. https://doi.org/10.31699/IJCPE.2023.2.14
16. Takacs G (2018) Electrical submersible pump components and their operational features. Electrical Submersible Pumps Manual: 55–152. https://doi.org/10.1016/B978-0-12-814570-8.00003-9
17. Barrios L, Prado MG (2009) Modeling two phase flow inside an electrical submersible pump stage. Proceedings of the International Conference on Offshore Mechanics and Arctic Engineering - OMAE 7:469–484. https://doi.org/10.1115/OMAE2009-79727
18. Zhu J, Guo X, Liang F, Zhang HQ (2017) Experimental study and mechanistic modeling of pressure surging in electrical submersible pump. J Nat Gas Sci Eng 45:625–636. https://doi.org/10.1016/J.JNGSE.2017.06.027
19. Beltur R, Prado M, Duran J, Pessoa R (2003). Analysis of experimental data of ESP performance under two-phase flow conditions. In Proceedings - SPE Production Operations Symposium, 463–473. https://doi.org/10.2118/80921-MS
20. Gamboa J, Prado M (2011) Review of electrical-submersible-pump surging correlation and models. SPE Prod Oper 26(4):314–324. https://doi.org/10.2118/140937-PA

21. Zhu H, Zhu J, Zhang HQ (2022) Mechanistic modeling of gas effect on multi-stage electrical submersible pump (ESP) performance with experimental validation. Chem Eng Sci 252: 117288. https://doi.org/10.1016/J.CES.2021.117288
22. Good P (2012) Ordinary least squares. A–Z of Error-Free Research:141–162. https://doi.org/10.1201/B12362-17
23. Martens HA, Dardenne P (1998) Validation and verification of regression in small data sets. Chemom Intell Lab Syst 44(1–2):99–121. https://doi.org/10.1016/S0169-7439(98)00167-1
24. Monte Verde W, Kindermann E, Bulgarelli NAV, Pastre LF, Foresti B, Bannwart AC (2024) A critical analysis and improvements of empirical models for predicting the performance of Electrical Submersible Pumps under viscous flow. Geoenergy Sci Eng 238:212871. https://doi.org/10.1016/J.GEOEN.2024.212871
25. Law M, Jackson D (2017) Residual plots for linear regression models with censored outcome data: A refined method for visualizing residual uncertainty. Commun Stat Simul Comput 46(4):3159–3171. https://doi.org/10.1080/03610918.2015.1076470
26. Zaheri MM, Ghodsi M, AlQallaf A (2018) Estimate of the head produced by electrical submersible pumps on gaseous petroleum fluids, a radial basis function network approach. Int J Artif Intell Appl 9(1):53–62. https://doi.org/10.5121/IJAIA.2018.9104

Vibration-Based Fault Diagnosis in Pipelines Using AI

Yaseen Al-Lawati, Mohammed Al-Kindi, Morteza Mohammadzaheri, and Loay Al-Lawati

Abstract This work presents a comprehensive simulation-based approach for detecting defects in steel pipelines using vibration analysis, combining two methodologies: linear regression and artificial neural networks (ANNs). In both studies, pipe models were developed in ANSYS, featuring 121–130 variations with single-hole defects at different longitudinal and circumferential positions, along with a defect-free control pipe. Modal analysis was conducted to extract the first 15 natural frequencies for each model, which were all below 5 kHz to ensure compatibility with real-world sensor systems.

In the first approach, a linear regression model was developed using MATLAB, where defect localization was based on the natural frequency differences between the control pipe and the defective models. The model achieved excellent performance across 70/30 and 80/20 training-test data splits, with R^2 values exceeding 0.998 and mean absolute percentage error (MAPE) below 5%. Generalization tests on unseen pipe configurations confirmed the robustness and reliability of the method.

In the second approach, three ANN architectures—feedforward, cascade-forward, and fitnet—were trained using the same frequency signature inputs. All models employed two hidden layers with 10 neurons each and were optimized using the Levenberg-Marquardt algorithm. Among them, the feedforward network showed the highest accuracy, especially in predicting fault positions on external test sets, with errors below 1% in some cases.

The combined findings demonstrate that both linear regression and ANN models can accurately identify defect locations in pipelines using modal frequency data. These results support the use of low-frequency vibration analysis in developing real-time, cost-effective, and non-destructive structural health monitoring (SHM) systems. Future work should focus on physical validation using embedded sensors to extend these promising simulation results into practical industrial applications.

Y. Al-Lawati · M. Al-Kindi · M. Mohammadzaheri (✉) · L. Al-Lawati
German University of Technology in Oman (GUtech), Halban, Oman
e-mail: Morteza.zaheri@gutech.edu.om

© The Author(s), under exclusive license to Springer Nature Singapore Pte Ltd. 2026
A. Azizi (ed.), *Hybrid Energy Systems and AI Integration*, Emerging Trends in Mechatronics, https://doi.org/10.1007/978-981-95-5859-9_4

Keywords Pipeline defect detection · Structural Health Monitoring (SHM) · Vibration analysis · Modal analysis · Artificial Neural Network (ANN) · Fault localization

Abbreviations

TET10	10-node Tetrahedral
AdaBoost	Adaptive Boosting
AI	Artificial Intelligence
ANN	Artificial Neural Network
AR	Auto-Regressive (feature)
BP	Back-Propagation
CFNN	Cascade-forward Neural Network
R^2	Coefficient of Determination
CNN	Convolutional Neural Network
ET	Eddy Current Testing
EMD	Empirical Mode Decomposition
FFT	Fast Fourier Transform
FPGA	Field-Programmable Gate Array
FRF	Frequency Response Function
GA	Genetic Algorithm
ID	Inner Diameter
LM	Levenberg-Marquardt
ML	Machine Learning
MAE	Mean Absolute Error
MAPE	Mean Absolute Percentage Error
MSE	Mean Square Error
MDNN	Mixture Density Neural Network
NDE	Non-Destructive Evaluation
NDT	Non-Destructive Testing
1DCNN	One-Dimensional Convolutional Neural Network
OD	Outer Diameter
RF	Random Forest
RMS	Root Mean Square
RMSE	Root Mean Squared Error
SVD	Singular-Value Decomposition
SHM	Structural Health Monitoring
SVM	Support Vector Machine
TDNN	Tapped Delay Neural Network
3D	Three Dimensional
UT	Ultrasonic Testing
VAR	Vector Auto-Regression (feature)
VAM	Vibro-Acoustic Modulation

Symbols

$\ddot{y}$	Acceleration
$\ddot{u}(t)$	Acceleration vector
σ_i	Activation Function
y_i	Actual Value
b_i	Bias Vector
$f_{control,\ i}$	Control Pipe Natural Frequency
$[C]$	Damping Matrix
ζ	Damping Ratio
f_{defect}	Defective Pipe Natural Frequency
μ	Damping (LM optimization parameter)
y	Displacement
$u(t)$	Displacement Vector
λ	Eigenvalue
${}_kS_i$	Fault Signature
f	Frequency
Hz	Hertz
$[M]$	Mass matrix
mm	Millimeter
$\varnothing$	Mode Shape Vector
ω_n	Natural Frequency
n	Number of Elements
$\widehat{y}_i$	Predicted Value
$[K]$	Stiffness Matrix
$\sum$	Summation Symbol
g	Training-gradient norm
$\overline{y}$	True Value
$\dot{y}$	Velocity
$\dot{u}(t)$	Velocity Vector
W_i	Weight Matrices

1 Section 1

1.1 Introduction

1.1.1 Background

Pipelines play a crucial role in transporting liquids and gases over long distances in the oil and gas, chemical, and water industries. They are central to international infrastructure due to their ability to move fluids efficiently and economically

[1]. However, pipelines are exposed to multiple threats, including corrosion, mechanical stress, and external damage, which can lead to leaks or structural failures. Such failures, if undetected, pose serious risks to environmental safety and operational continuity [2].

Traditionally, non-destructive testing (NDT) methods such as ultrasonic testing (UT), magnetic flux leakage, and radiographic inspection have been widely used to detect pipeline faults. Despite their effectiveness, these methods involve scheduled inspections, require significant labor, and are unsuitable for buried or inaccessible pipelines [3, 4]. Moreover, they are generally incapable of providing real-time feedback or continuous monitoring.

Vibration-based structural health monitoring (SHM) techniques offer an attractive alternative. These methods detect structural anomalies by analyzing shifts in vibration patterns, such as changes in natural frequencies or mode shapes [5]. Vibration-based techniques are non-invasive, cost-effective, and capable of remote monitoring, making them ideal for large or inaccessible pipeline networks.

Despite their advantages, vibration-based methods face operational and technical limitations, including noise sensitivity, signal complexity, and the need for expert interpretation [6, 7]. To address these challenges, artificial intelligence (AI), particularly machine learning (ML), has been introduced into SHM systems. ML models, such as artificial neural networks (ANNs), can identify subtle relationships in vibration signals and learning patterns that correspond to defect characteristics [8, 9].

This study builds on these developments by combining two data-driven approaches—linear regression and ANN-based learning—for the detection and localization of defects using modal frequency data from ANSYS simulations.

1.1.2 Problem Statement and Research Gaps

Over time, pipelines are exposed to various degradation mechanisms, which affect their structural integrity and pose significant safety and environmental concerns [2]. Traditional inspection methods such as UT and magnetic flux leakage require direct access, are labor-intensive, and are often insufficient for real-time fault diagnosis in complex or buried pipelines [4, 10].

Although vibration-based SHM has emerged as a viable option, it presents challenges. Signal interpretation remains complex due to environmental noise and operational variations [11]. Many studies rely heavily on simulation data, lacking validation with field-based testing [10]. Moreover, current ANN models are often treated as black-box systems, making their internal decision-making difficult to interpret, especially in safety-critical applications [6, 12].

Furthermore, the limited availability of labeled vibration data and the need for high computational resources reduce the practicality of ML-based SHM systems [1, 13]. Linear regression, while more interpretable, may not capture nonlinear fault features. Therefore, there is a need to systematically evaluate and compare both models using a standardized dataset and external testing to measure generalization and practical usability.

This research aims to address these gaps by building both linear regression and ANN models based on a common vibration dataset extracted from ANSYS simulations of defective pipelines and evaluating their performance and accuracy in both internal and external test conditions.

1.1.3 Objectives and Scope

The core objective of this study is to develop and compare the effectiveness of linear regression and artificial neural network (ANN) models in diagnosing pipeline faults using modal frequency shifts. The project relies on simulations conducted in ANSYS to generate vibration data across various defective pipe scenarios.

Specifically, the goals include:

- Designing 121–130 steel pipe models, each featuring a single hole at different longitudinal and circumferential positions.
- Performing modal analysis to extract the first 15 natural frequencies, all below 5 kHz for practical sensor compatibility.
- Generating input features by subtracting the control pipe's frequency vector from each defective model's vector.
- Training a MATLAB-based linear regression model and testing it using both 70/30 and 80/20 data splits.
- Designing three ANN models—feedforward, cascade-forward, and fitnet—with two hidden layers (10 neurons each), using the Levenberg-Marquardt algorithm
- Evaluating all models using error metrics: MAE, RMSE, MAPE, and R^2.
- Performing external validation using nine previously unseen pipe configurations to test generalization.

While this research is limited to simulated conditions, it lays the foundation for future real-world implementation. The findings are expected to support the integration of AI models with vibration-based fault detection in SHM systems.

1.1.4 Structural Health Monitoring in Pipelines

Importance of SHM in Pipeline Infrastructure

SHM stands as an essential infrastructure management tool for the assessment of structural conditions like the integrity of bridges, buildings, and pipelines. SHM systems within the pipeline field play an essential role in detecting defects at an early stage, preventing a catastrophic system failure from occurring. During the past twenty years, the field of SHM has improved significantly through the advancements of sensing technologies, signal processing techniques, and data acquisition systems [14, 15].

Techniques for Structural Assessment

Pipeline monitoring techniques based on SHM have different capabilities and operational boundaries. One of the vibration-based techniques used for detecting damage is modal analysis, which is done by seeing the changes in natural frequencies, the damping ratios, and mode shapes of the pipe [16]. These parameters can detect structural changes and can show early indicators of anomalies. The implementation of modal analysis faces challenges because it requires advanced experimental equipment together with controlled excitation [7].

Guided Ultrasonic Wave Methods

Guided ultrasonic waves have become a method used for inspecting long-distance pipelines. These ultrasonic waves can travel long distances within a pipeline network, where they can detect changes like corrosion or cracks [17]. Torsional waves have demonstrated successful applications in locating pipeline damage without the need for a complete surface exposure. However, these guided wave signals can have some issues related to reflections, mode conversions, and noise, especially under variable environmental conditions [5, 18].

Sensor Networks and Data Acquisition Systems

The effectiveness of structural health monitoring (SHM) systems relies heavily on the integration of sensor networks and data acquisition systems. Research demonstrates that sensor network design and the sensor's density influence the sensitivity and precision of the monitoring systems. Research about wireless sensor networks (WSNs) has been increasing due to their ease of deployment and their ability to operate in complex environments. However, some challenges exist with these wireless sensor networks related to power management systems, transmission reliability, and latency issues [19]. A reprogrammable integrated circuit called a field-programmable gate array (FPGA) can be integrated into wireless sensor networks to boost system reliability and efficiency. FPGAs operate efficiently in embedded SHM systems, which detect pipeline defects under harsh conditions because they combine adaptable abilities with powerful computing capabilities [15].

Integration of Artificial Intelligence in SHM

The adoption of machine learning together with artificial intelligence within SHM systems creates new ways of detecting damage. AI-enhanced SHM systems at present incorporate historical data to learn behavioral patterns that guide their predictive activities. Artificial intelligence systems in modern times reduce the need for human interpretation while simultaneously improving their fault detection abilities [14].

Operational Challenges and Deployment Limitations

Multiple issues must be addressed before SHM operations can be successfully applied to pipelines. Detection systems encounter several problems during operation due to the environmental conditions, sensor wear, and noise-generated disruptions. Additionally, the high costs of deployment and maintenance expenses affect the use of SHM systems in extensive pipeline networks [20, 21].

1.1.5 Vibration-Based Fault Detection Techniques

Principles of Vibration-Based Fault Detection

Vibration-based fault detection serves as a technique in structural health monitoring (SHM) systems. Structural defects such as cracks, corrosion, or leaks can cause changes in the pipeline's normal vibration behavior. This change can be studied and analyzed to indicate the presence of an anomaly [22]. This method detects the natural frequency shifts, changes in the damping ratios, and deviations in mode shapes. The detection can be made by placing vibration sensors along the pipeline according to [23].

Signal Processing Techniques in Fault Detection

Modal analysis, wavelet transformation, and frequency response functions (FRF) have proven their effectiveness in detecting pipeline faults. The pipeline defect detection accuracy achieved by systems that utilized FRF methods reached within 2% of actual values with noise disturbances, according to [24]. Results from the detection process rely on choosing the best sensor positions along pipeline pathways.

The research by Razi and Taheri [25] proved that empirical mode decomposition (EMD) works well for submarine pipeline inspections since it detected cracks with 99% accuracy at 0.1% error precision. Chirp excitation methods proved to be better compared to traditional techniques to detect small cracks, as they produced more reliable outcomes.

Environmental and Operational Challenges

Real-world applications can bring challenges to vibration-based fault detection systems. Saidin et al. [11] found that their system generated 4.5% of incorrect positive results because of environmental disturbances and changes to the operational environments. Their research indicated that the detection systems loses between 8% and 12% sensitivity when operating under changing environmental factors; in this way, the monitoring reliability decreases within demanding operational spaces.

Ravizza et al. [7] noted that noise interference presented a vital system limitation that degraded signal clarity between 15% and 18% when the system started processing over 10,000 signals. Their efficiency decreased because of the classification errors that reached up to 8% due to the growing computational complexity. To ensure the effectiveness of vibration-based systems in large-scale pipelines, advanced signal processing methods together with noise-reduction techniques are needed.

Comparison with Traditional NDT Methods

It is necessary to have a complete understanding of the strength points and limitation points of vibration-based methods as compared to ultrasonic testing (UT) and eddy current testing (ET). Although vibration-based SHM can assess the overall pipe's health, it cannot identify small defects. UT and ET are still a very reliable option for inspecting small areas, where they can spot inner cracks or thinning in the material. With high-resolution defect detection, they are an excellent choice for carrying out maintenance on targeted areas [26].

In particular, vibro-acoustic modulation (VAM) has shown great ability to detect tiny cracks. Researcher Golchinfar [26] indicated that VAM was 30%–35% more sensitive than both UT and ET. Even so, the study found that about 5% of positive results could be false. Vibration-based methods can offer an affordable and flexible choice. However, using UT or ET in addition to that can raise the fault detection rate and its accuracy.

Integration with Advanced Models

The detection of equipment faults through vibration analysis has reached higher efficiency levels in recent years. According to Zajam et al. [27], wavelet analysis working with machine learning models has raised defect detection accuracy to 87%–90%. This was done by the fusion ANNs and deep learning models. The process has become automated, which minimizes human involvement for defect classification while improving the system's reliability in fault detection.

Table 1 provides a comparison between different vibration detection methods for SHM systems. The table presents an overview of fault detection methods which includes their performance advantages and their limitation

The detection of pipeline faults through vibration analysis proves to be a dependable approach, as studies demonstrate both high accuracy levels exceeding 99% and minimal error ranges at only 0.1% for the pipeline. However, challenges such as

Table 1 Comparative overview of key studies on vibration fault detection

Study	Technique used	Strengths	Limitations/Challenges
[22]	Vibration-based discriminant analysis	Classified leak conditions with up to 95% accuracy	Relies on vibration patterns; not frequency spectrum
[23]	Sensor-based vibration monitoring (frequency, damping, mode shapes)	Enables early fault detection via parameter shifts	Requires optimal sensor placement
[24]	Frequency Response Function (FRF)	Detection accuracy within 2% despite noise	Dependent on noise levels and sensor layout
[25]	Empirical Mode Decomposition (EMD)	99% crack detection accuracy, 0.1% error	Focused on submarine pipelines; method-specific scope
[11]	Vibration analysis under environmental changes	Adaptability noted; performance tested in varied environments	False positives at 4.5%; 8%–12% sensitivity drop in harsh conditions
[7]	Signal processing under noise and large datasets	Supports large-scale analysis; highlighted need for advanced processing	Signal clarity loss 15%–18%; 8% error rise due to computational complexity
[27]	Wavelet + Machine Learning (ANNs, Deep Learning)	87%–90% accuracy; automation reduces human input	Noted computational demands; generalization challenges

external noises decrease the accuracy by (15%–18%), highlighting areas for improvement. The fault detection abilities of vibration-based detection methods can be effective but integrating it with other NDT solutions could deliver an optimal pipeline monitoring system.

1.1.6 Integration of Artificial Intelligence and its Subfield Machine Learning (ML) in SHM

Role of Machine Learning within Artificial Intelligence for Fault Detection

AI —particularly ML— technologies have brought revolutionary changes to the fault detection practices by instituting automatic structure data examination systems for faster identification of abnormalities within certain structures [28]. Structural health monitoring (SHM) benefits from artificial intelligence models, especially ML algorithms, because they enhance the reliability in detecting faults, along with their precise location. Historical training vibrational data in machine learning algorithms helps them detect specific defect patterns, which enables quick detection of defects beyond traditional human capabilities [8]

Machine learning techniques such as support vector machines (SVMs), artificial neural networks (ANNs), random forest (RF), and convolutional neural networks (CNNs) are commonly used to analyze data. The complex vibration patterns and large datasets are easier to handle through these models [29, 30].

Applications and Performance of ML Models

Rajasekaran et al. [31] built a pipeline leak detection system that combined 1D convolutional neural networks (1DCNN) and an adjusted AdaBoost algorithm model, where the model achieved 97.4% success in detecting oil and gas pipeline leaks. Another study that combined wavelet analysis with an SVM classifier employed by [27] achieved an 86.2% training accuracy, followed by testing accuracy between 87% and 90% in detecting the defected region.

Another article by Singh et al. [13] showed that artificial intelligence models provide versatile capabilities for finding defects, highlighting that machine learning techniques can adjust easily to various kinds of weaknesses and operating conditions. The researchers proved that supervised learning models operate successfully while maintaining high accuracy results. These high-accuracy results are achieved by using time-series data from vibration sensors as input for these models, which later on can be executed precisely in structural analysis with high precision.

Limitations and Practical Challenges

AI models used for SHM systems face questions about their reliability. A vital concern, according to Leite et al. [12], is that black-box AI models like deep neural networks, which achieve high accuracy yet remain unclear regarding their decision-making procedures. The lack of interpretability in AI models prevents engineers, unfamiliar with data-driven modelling and validation, from relying on their output, particularly in safety-critical systems.

ML training faces its greatest challenge in obtaining quality and quantity of labeled training data. The accuracy rates of the systems decrease and false positive rates increase due to insufficient training data, according to Singh et al. [13] during the analysis of complex and uncommon defect scenarios. Moreover, models trained on data from one pipeline system may not be the same from one pipe to another, resulting in performance decline when new environments are encountered.

Wong & McCann [1] found that distributed sensor network data analysis requires high computational complexity because of its large dataset volume. The processing of real-time vibration data for edge-based SHM systems presents operational difficulties for low-power systems.

The reviewed studies utilize a wide range of ML techniques within AI for vibration-based fault detection, but hybrid and deep learning models prove to be the most effective according to Table 2. Different performance measurement standards across various studies hinder the ability to properly assess the effectiveness of their results on equivalent scales.

Research papers demonstrate that ML models improve SHM system efficiency in fault detection accuracy by achieving detection rates between 86.2% and 97.4%. The large-scale application of AI-based fault detection systems faces three major obstacles: limited training data, computational complexity, and lack of interpretability. The solution to these problems is further studies and investigations that aim to

Table 2 Comparison of ML techniques within AI for fault detection

Study	AI/ML Technique	Application	Key Strengths	Limitations
Rajasekaran et al. [31]	1D CNN + Modified AdaBoost (SVM, DT, MLP)	Water pipeline leak detection and localization	High precision and hybrid model effectiveness	Still needs testing in real-world environments
Pan et al. [29]	SVM with feature extraction (AR, VAR, SVD)	Structural damage detection in civil structures	Feature extraction enhances SVM classification	Complex setup and sensitivity to feature choice
Zhang et al. [30]	CNN with Phase-based Motion Estimation	Structural damage localization in steel frames	Low training data needs; high fault localization	High dependence on video quality and setup
Malekloo et al. [8]	General ML (overview of SHM integration)	Big data and ML in SHM across domains	Covers cutting-edge technologies	General overview; lacks case-specific insights

improve the model stability, minimize false positives, and boost performance in complex environments.

1.1.7 Application of Artificial Neural Networks (ANNs) in Vibration-Based Pipeline Monitoring

Application of ANNs in SHM and Pipeline Monitoring

ANNs can detect complex and nonlinear patterns in vibration signals, making them a suitable option for automated structural health monitoring (SHM) [32]. The training of ANNs for pipeline monitoring systems enables them to identify and categorize pipeline anomalies through analysis of minor variations in time-series sensor data. After training their systems, they can perform rapid evaluations of pipeline health conditions, becoming an effective application for continuous tracking [9].

The pattern recognition abilities of ANNs are compatible with vibration-based monitoring because they can learn from time-domain features such as root mean square (RMS), standard deviation, and peak values to identify the structural states. Research by Samal et al. [33] revealed that ANN models operating on faulty mechanical vibration signals reached classification results exceeding 95% accuracy while functioning in noisy conditions.

Performance Across Case Studies

Multiple research studies have implemented ANN models directly for pipeline monitoring systems, which have proved successful. The research by Barradas et al. [34] showcased that tapped delay neural networks (TDNN) can locate and detect multiple leaks by processing two pressure measurements alone, proving that proper ANN training can produce results from limited data. Under complex multiphase flow conditions, Al-Ammari & Sleiti [35] created a leak detection model was created based on ANN. The results show that the localization error was below 1% for single leaks while maintaining under 8% accuracy for multiple leaks, thus proving that ANN works in real-world dynamic conditions. Khan et al. [36] used mixture density neural networks (MDNNs) to develop a cost-efficient failure detection system for water distribution systems. The study demonstrated that ANNs can maintain their operational quality with minimum resources.

The study by Karabacak [32] demonstrated that ANN models delivered better results compared to SVMs during fault diagnosis of internal combustion engines based on vibration data where accuracy reached 97% while SVMs achieved 95.8%. The system evaluation conducted was not pipeline-based, yet it used similar methods and signal types to traditional SHM applications, which makes its performance benchmarking reasonable.

Challenges and Limitations of ANN Models

Many problems affect various machine learning models that appear most intensely within ANNs. Such as the high need for data and its risk of overfitting. It needs a large dataset that contains extensive information to achieve good generalization results. The collection of datasets in SHM applications can become complicated because of variability in fault types, structural responses, and environmental conditions. It is trained through limited data may focus on memorizing noise or outliers instead of recognizing generalizable patterns [6, 37].

Real-time monitoring faces significant obstacles because of the high computational complexity that exists in ANNs. The deep architectural design requires large amounts of memory storage and an incredible amount of processing capabilities, which makes it difficult to deploy on embedded systems or remote locations with restricted hardware resources [6, 9].

The accuracy of ANN models depends heavily on the precise quality combined with complete input data. Its predictions can suffer from low accuracy when either sensors fail to operate correctly, or when signals become noisy, or when any data entry is missing. The methods of dropout training and data denoising are invoked to increase model robustness according to [37]. Implementing these methods could lead to more complicated models and might not fully fix the core problems.

Different ANN-based approaches that apply to pipeline fault detection and structural health monitoring can be seen in Table 3. Throughout the analysis, numerous ANN applications become evident while exposing the common technical hurdles that need resolution for dependable actual implementation.

Artificial neural networks remain a method in vibration-based pipeline monitoring because they provide exceptional accuracy, together with high flexibility. The practical application of these models requires dealing with increased data needs together with increased computational requirements while experiencing reduced interpretability of results. Its development requires attention to system transparency

Table 3 Comparison of artificial neural network (ANN) applications in vibration-based SHM

Study	Application	Strengths	Limitations
Karabacak [32]	Fault detection in engines using vibration signals	ANN outperformed SVM with 97% accuracy	Tested on engines, not pipelines
Mohamed et al. [9]	SHM pipeline safety assessment with AI	AI enables fast, automated pipeline assessments	Computational complexity and interpretability issues
Samal et al. [33]	Fault classification in noisy vibration data	High classification accuracy (>95%) in noisy environments	No details on ANN architecture or scalability
Barradas et al. [34]	Leak detection in pipelines using two pressure inputs	Effective leak detection from minimal data	Does not address multi-type leak environments
Al-Ammari and Sleiti [35]	Leak detection in complex multiphase flows	Localization error <1% for single leaks	Performance drops for multiple leaks (≤8% accuracy)
Khan et al. [36]	Low-cost failure detection for water systems	Cost-effective, minimal-resource ANN system	Early model; not tested on large-scale systems
Azanaw [6], Deng et al. [37]	Challenges: data quality, overfitting, noise, interpretability	Identifies ANN limitations in SHM (black-box, data dependency)	Real-time deployment hindered by hardware

and data diversity, along with hardware limitations due to the challenges this technology presents.

1.1.8 Research Gaps

Even though previous SHM research papers are aware of the same obstacles that appear in using ANN, such as small datasets, the use of lab environments, and limited computational optimization, important gaps are still present, mainly about using vibrations to locate faults in pipelines [23, 28].

Research about vibration-based SHM for pipeline structures has not been studied thoroughly. Most of the research paper focuses on bridges and rotating machinery, rather than studying how pipelines vibrate differently under various conditions [32, 38]. As a result, current approaches are limited in their ability to be applied to actual pipeline inspection.

Most existing studies suggest methods for detecting but not locating them, few studies investigate how the position of the defect along a pipe can affect the modal frequencies, which can then be used to locate it. The absence of detailed simulations or experimental mapping of defects at various locations makes it difficult to compare defect locations and their vibrational signatures [24, 27].

Although ANNs are widely used in SHM studies, they have not been tested in a real-life scenario. Researchers often work with small, clean data collections without any noise. Challenges such as dataset detail, overfitting, and model generalization are often overlooked and focused on the results only [28, 32]. Currently, there are few studies where ANN models are tested using common error measurements to check how well they can find the exact location of the fault.

To fill those gaps, this research uses simulation-based modal frequencies from 130 pipes, 129 with localized defects and one control pipe, to assess and compare three ANN architectures along with a linear regression model. The system was built to detect defects in the longitudinal direction by examining the natural frequency changes obtained through modal analysis. The suggested framework justifies the use of the simulation, which then can be a starting point for future real-world SHM implementation.

2 Section 2

2.1 Methodology

2.1.1 Methodological Framework and Research Design

The objective of the research is to create a method using vibrational analysis along with ANN and linear regression for locating defects in steel pipelines through simulation. Frequencies below 5 kHz will be analyzed only. The research demonstrates that low frequencies can detect structural defects in a system. This approach can be used instead of traditional methods for ensuring quality in manufacturing and inspection services.

Finite Element Modeling

There were 130 pipes modeled in Ansys SpaceClaim, among which one was perfect and the other 129 had defects. For the faulty models, each one had a hole that was found in 15 different positions along the length and within eight different axes. The axes are distributed every 45° around the pipe's cross-section. All 15 holes were evenly spaced out at a distance of 32.72 mm. The pipe has a total length of 1000 mm, an outer diameter of 33 mm, and an inner diameter of 30 mm. Boundary conditions were set at the start and end of the pipe to represent clamps, these conditions were only set over a 25 mm region on both ends. Around 9 modeled pipes were set apart for the external test and modeled the defect positions differently than the rest of the pipe models.

Modal Analysis

The first 15 modes of natural frequencies were found from each pipe model using modal analysis, where frequencies simulated were under 3500 Hz. The pipes were simulated in ANSYS (Student Version 24.2).

Fault Signature Comparison

For every defect model, its natural frequency was subtracted from the control pipe natural frequency:

$$ {}_kS_i = f_{defect,i} - f_{control,i}, \quad i = 1, 2, \ldots 15. \tag{1} $$

In this way, we have produced 15 unique elements for each hole and named them as fault signature. This signature will act as an input to our machine learning model. But since we are only detecting the location of the defect, this produces 120 signature elements for each position.

Neural Network-Based and Linear Regression Localization

The signatures of the faults were calculated to train regression-based artificial neural networks (ANNs) for predicting the location of the defect in millimeters. Results are based on using MATLAB neural architectures like feedforward, cascade-forward, and fitnet with 10 neurons used in each hidden layer [10 10]. The choice of two hidden layers resulted from a trial-and-error approach and comparing each architecture with the MSE and MAE, which is detailed in the result section. A total of 120 samples made up the dataset, where each sample had 15 "frequency shift" features and a continuous "defect position" output. In addition to ANN, a linear regression model was also developed using the same input-output structure and dataset. Two versions of the regression model were trained: Model A used a 70/30 training-to-testing split, and model B used an 80/20 split. These models provided a baseline for evaluating performance and interpretability, and their results are presented alongside the ANN models in the Results section.

Model performance was assessed using standard regression metrics:

- Mean absolute error (MAE):

$$ MAE = \frac{1}{n}\sum_{i=1}^{n} | y_i - \widehat{y}_i |. \tag{2} $$

- Root mean squared error (RMSE):

$$ RMSE = \sqrt{\frac{1}{n}\sum_{i=1}^{n} (y_i - \widehat{y}_i)^2}. \tag{3} $$

- Coefficient of determination (R^2):

$$R^2 = 1 - \frac{SS_{res}}{SS_{tot}}. \quad (4)$$

- Mean absolute percentage error (MAPE):

$$MAPE = \frac{100\%}{n} \sum_{i=1}^{n} \left| \frac{y_i - \widehat{y}_i}{y_i} \right|. \quad (5)$$

In these Eqns. (2), (3), (4), and (5), y_i represents the actual true value, $\widehat{y}_i$ is the predicted value, n is the total number of observations, and $\sum$denotes the summation over all observation from $i = 1$ to n. SS_{res} and SS_{tot} are the residual sum of squares and total sum of squares respectively.

Unfortunately, due to technical and accessibility issues, it was difficult to do the experiment part using a high-resolution displacement sensor to validate our simulation data. Initially, the experiment was planned to be done by applying an impulse to the pipe and capturing its displacement over time. However, considering the damping ratio of mild steel $\zeta = 0.01$, limits the usable vibration duration. According to the second law linear differential equation for damped motion:

$$\ddot{y} + 2\zeta\omega_n\dot{y} + {\omega_n}^2 y = 0. \quad (6)$$

where $\ddot{y}$ is the acceleration, $\dot{y}$ is the velocity, y is the displacement, ζ (zeta) is the damping ratio, and ω_n is the natural angular frequency of the system.

It was found that the vibration would die out within 26 ms after an impulse, leaving only 260 time-domain samples available for analysis. 10,000 samples are required for effective ANN training. As a result, modal analysis was an alternative option since its implementation offered an efficient solution and allowed for scalability.

Figure 1 illustrates a simulated displacement vs time graph for an impulse of 250 N for 1 ms at the center of the pipe with a mass of 1 kg and fixed on both ends. The figure represents the limitation of the time-domain samples' availability; therefore, modal analysis was used instead.

To ensure the natural frequencies were obtained correctly, five simulations were repeated for each model to have a reliable set of data. Figure 2 illustrates the sequence from simulation to frequency extraction and the application of an ANN. The section that follows explains the geometric modeling, meshing techniques, and defect parameterization used in ANSYS.

2.1.2 Pipe Model and Geometric Configuration

All the pipes were modeled as a straight hollow cylinder in SpaceClaim Ansys to mimic the steel pipeline in real life. The total length of the pipe was 1000 mm with an outer diameter of 33 mm and 30 mm as an inner diameter with a uniform thickness of

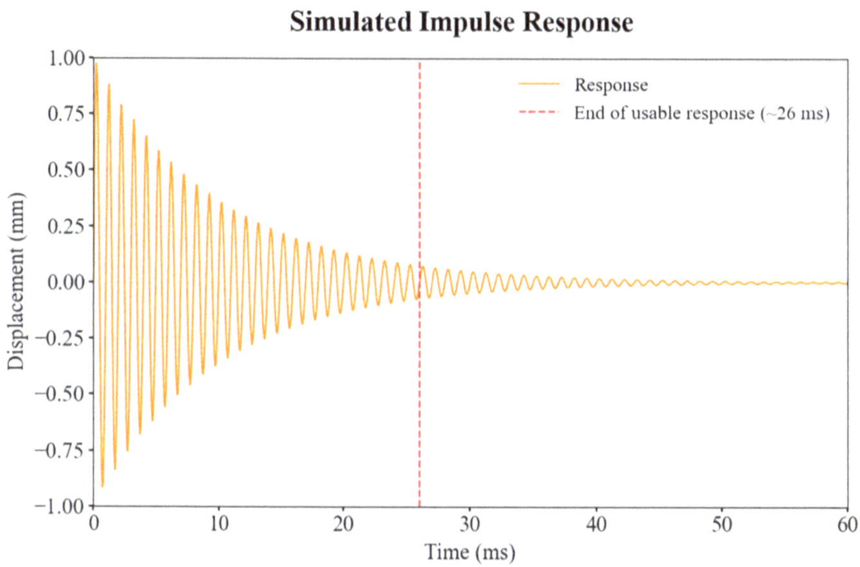

Fig. 1 Simulated impulse response of pipe to a 250 N force

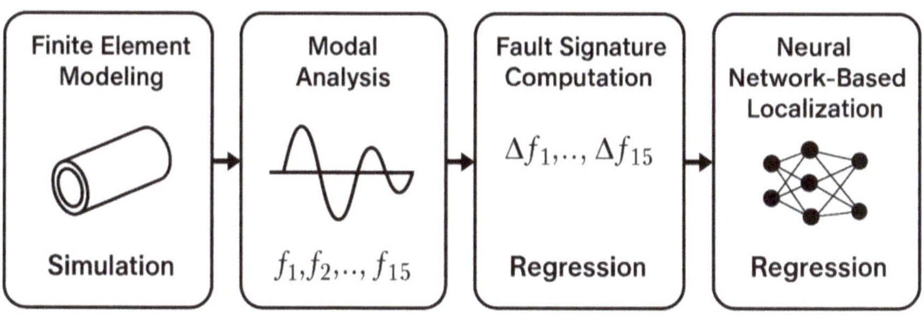

Fig. 2 Workflow of the vibration-based pipe fault detection framework

3 mm. 25 mm at both ends are selected to be fixed. These dimensions were selected to match the real experiment that was planned earlier but was not conducted due to laboratory restrictions to validate the simulated data. The pipe specification can be seen in Fig. 3 and Table 4

Defects were made with an 8 mm circular hole piercing completely through the pipe walls, simulating damage that can be caused by a drill. These defects were chosen to match what we would have done in our physical experiment. The hole geometry was created by sketching a circle on a plane and then projecting it onto the pipe outer surface. With the use of the pull command, the hole was created. Figure 4 showcase the process of creating the holes.

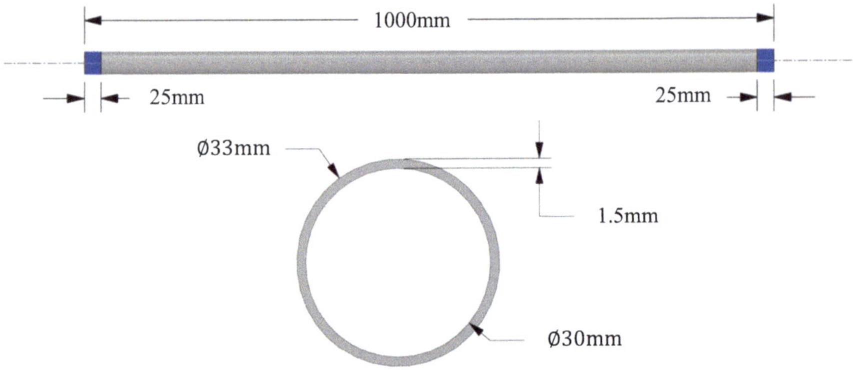

Fig. 3 Front and side view of the pipe

Table 4 Pipe dimension and material properties

Property	Value	Unit
Pipe length	1000	mm
Outer diameter (OD)	33	mm
Inner diameter (ID)	30	mm
Wall thickness	3	mm
Fixed end length	25 (each)	mm
Material	Mild steel	—
Young's modulus	210	GPa
Density	7850	kg/m^3

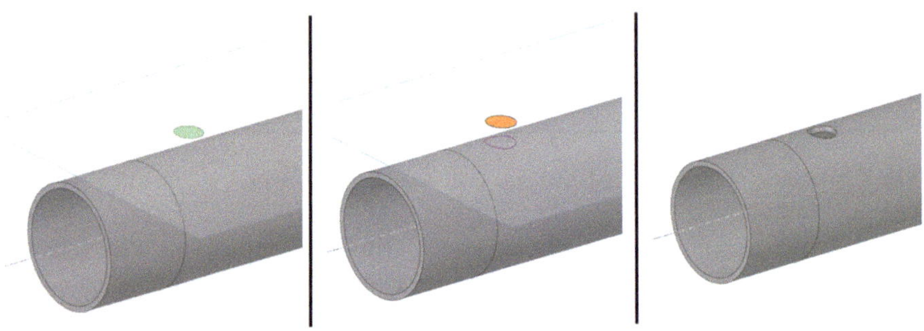

Fig. 4 Overall process of defect modeling

A total of 120 defective pipe models were created, each having a single hole at a targeted position across 15 locations and 8 circumferential axes. These holes were placed using half the pipe's length, based on symmetry, with each having a separation of 32.72 mm. These positions were measured from the end of the fixed face to the center of the pipe. Figure 5 shows how all the holes were distributed in the 120 models.

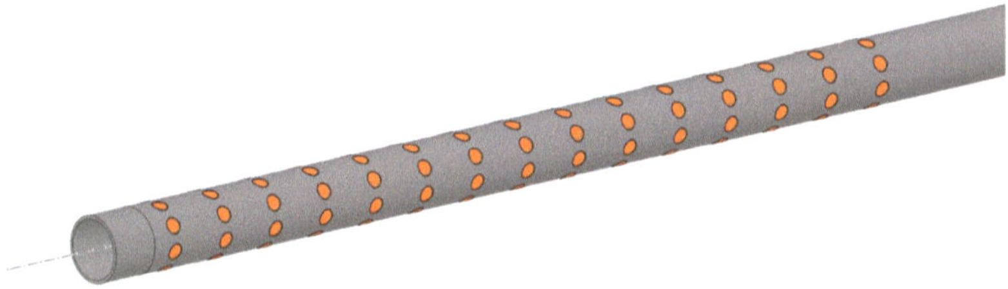

Fig. 5 Defect layout: 15 longitudinal positions × 8 circumferential axes

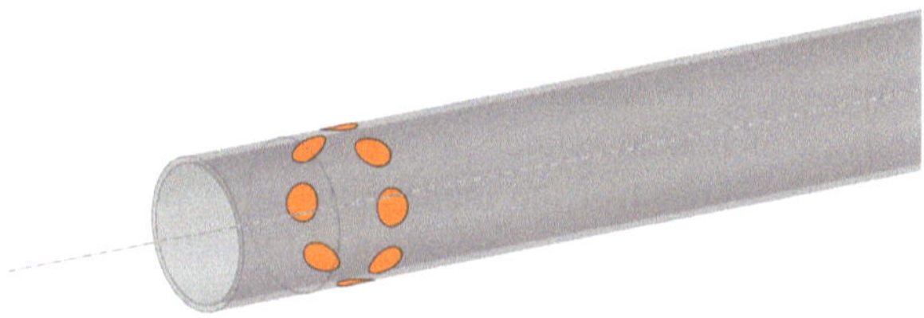

Fig. 6 Defect layout: 8 circumferential axes

Table 5 Defect position index: longitudinal distances

Position index	Distance from fixed end (mm)
1	39.00
2	71.72
3	104.44
4	137.16
5	169.88
6	202.60
7	235.32
8	268.04
9	300.76
10	333.48
11	366.20
12	398.92
13	431.64
14	464.36
15	497.08

The holes around the outer surface pipes were distributed evenly at a 45° interval covering the pipe's circular cross-section, i.e.: 0°, 45°, 90°, . . ., up to 315°. This allowed for full coverage of the pipe in case a defect appears around its circumferential axis (Fig. 6).

Table 5 provides the positions selected for the holes to be modeled on; the measurements were taken from the left end of the hole, starting from 39 mm and ending with 497.08 mm. To test the effectiveness of the neural network, nine additional models were created to test its accuracy on new data that it hadn't been trained on before. New nine unique positions were chosen with different axes. The following subsections describe the meshing strategy and boundary condition setup used to prepare the models for modal analysis.

2.1.3 Mesh and Boundary Condition Setup

Mesh Generation

All the pipes were meshed using ANSYS Mechanical software, with automatic meshing enabled. 10-node tetrahedral elements (TET10) were applied to the geometry, which are second-order elements intended to correctly capture the geometric curves, such as in cylindrical pipes and round defects. These elements allowed for the improvement of the modal simulation precision by utilizing mid-side nodes, resulting in a better representation of the deformation and strain gradients [39].

A 5-mm global element size was chosen to be applied to all 130 models. No manual local refinements or inflation layers were introduced, as the study focuses on the overall vibrational behavior. It applied an adaptive sizing, a fast transition method, and medium soothing. The resolution level was set to default (2). Figure 7 shows the TET10 mesh on the control pipe body (3D view) and Fig. 8 shows a closer look to mesh generated around the defect hole.

Mesh Sensitivity Analysis

To see whether the mesh size was indeed a good option, a mesh convergence study was conducted. Two cases were used to analyze:

1. The control pipe (defect-free)
2. A defective pipe model with a single hole located at 39 mm (first unique position out of the 15) from the left end and oriented at 0°.

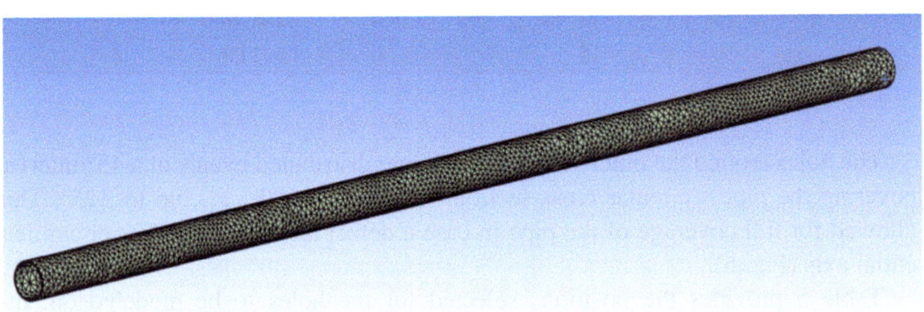

Fig. 7 Mesh view of control pipe

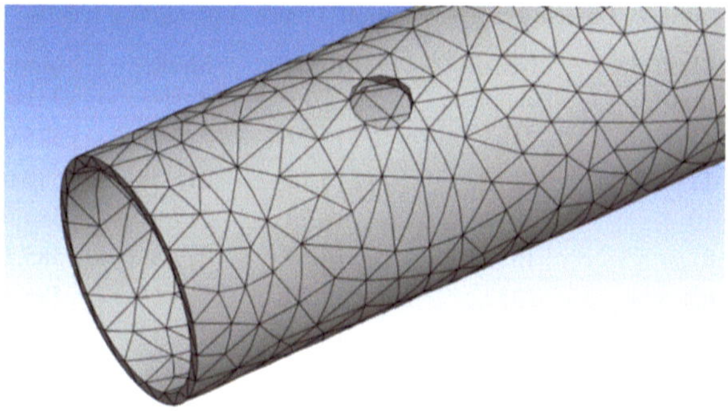

Fig. 8 Hole mesh (closer view)

Table 6 Control pipe mesh convergence results

Mesh size (mm)	Element Size	Mode 1 (Hz)	Mode 2 (Hz)	Mode 3 (Hz)	Mode 4 (Hz)	Mode 5 (Hz)
15	8132	218.36	218.46	589.65	589.96	1126.5
12	8322	218.24	218.3	589.36	589.53	1126
10	11355	218.09	218.09	588.95	588.95	1125.2
8	15935	218.01	218.01	588.74	588.74	1124.8
6.5	22563	217.96	217.96	588.56	588.56	1124.3
5	23980	217.9	217.9	588.41	588.42	1124.1
4	38790	217.9	217.9	588.41	588.41	1124.1
3.5	49137	217.82	217.82	588.2	588.2	1123.7
3	65339	217.85	217.85	588.27	588.27	1123.8
2.5	97369	217.81	217.81	588.18	588.18	1123.77
2	149502	217.78	217.78	588.17	588.17	1123.77
1.5	267447	217.76	217.76	588.12	588.12	1123.72
1	915345	217.76	217.76	588.1	588.1	1123.71

For each case, a sizing between 15 mm and 1 mm was analyzed, taking the first five natural frequencies to compare them with different mesh sizes, as can be seen in Tables 6 and 7.

Figures 9, 10, 11, 12, 13 represent the plot graphs of the first five modes and how the frequencies change over the mesh size. The graph shows that the natural frequency settles into a steady value at smaller mesh sizes, that doesn't change much with further refinement. In contrast, for larger mesh sizes meshes, it causes the predicted frequency to increase, hinting that they may not capture the pipe's behavior as accurately (Figs. 14, 15, 16, 17, 18).

To summarize the mesh-convergence study on both the control pipe and a pipe with a hole showed that using 5 mm element size yields modal frequencies within 0.1–0.3 % of those obtained with much finer meshes (e.g., 3.5 mm or 1 mm), while keeping element counts to around 24,000 per model. Adjusting the mesh to be smaller results in doubling the number of elements and therefore increases the

Table 7 Defected pipe (39 mm, 0°) mesh convergence results

Mesh size (mm)	No. of elements	Mode 1 (Hz)	Mode 2 (Hz)	Mode 3 (Hz)	Mode 4 (Hz)	Mode 5 (Hz)
15	7103	212.27	213.93	575.98	579.32	1102.7
12	8340	215.12	215.58	580.66	581.34	1111.6
10	10840	216.73	217.9	584.95	587.69	1119.6
8	15112	216.89	217.8	585.13	588.36	1119.1
6.5	23522	217.09	218.14	586.58	588.93	1121.3
5	24037	217.3	218.32	587.05	589.36	1122
4	38166	216.96	218	586.23	588.56	1120
3.5	49386	216.9	217.98	586.1	588.52	1120.4
3	64664	216.86	217.95	585.97	588.42	1120.1
2.5	96964	216.81	217.93	585.84	588.36	1120.3
2	154288	216.76	217.90	585.71	588.29	1120.3
1.5	271358	216.71	217.88	585.58	588.22	1120.4
1	957350	216.66	217.85	585.45	588.15	1120.4

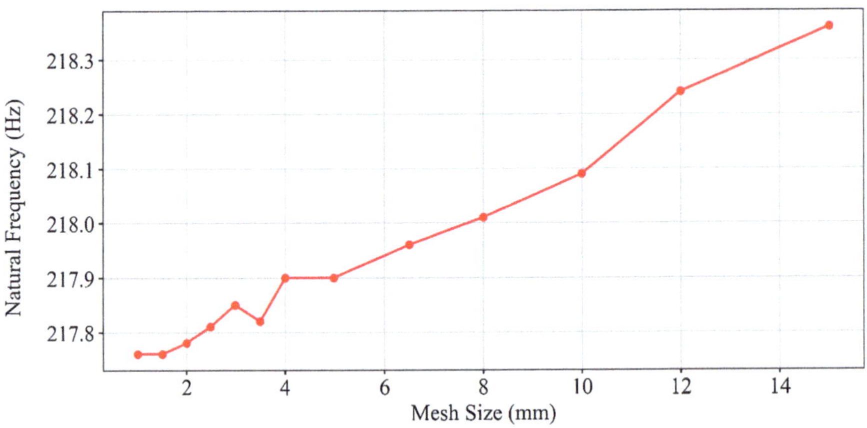

Fig. 9 Control pipe-mesh convergence (Mode 1)

solving time and memory usage with only producing small marginal improvements in the first five natural frequencies. Given that 130 pipe geometries needed to be analyzed, the 5 mm mesh provided the optimal balance between modal-frequency accuracy and computational efficiency.

Boundary Conditions

Both ends of the pipe and a 25 mm section were designated to act as fixed support. It was applied to the surface constraints at all six degrees of freedom (UX, UY, UZ, ROTX, ROTY, ROTZ). This was done to simulate fully clamped ends and was done for all the models.

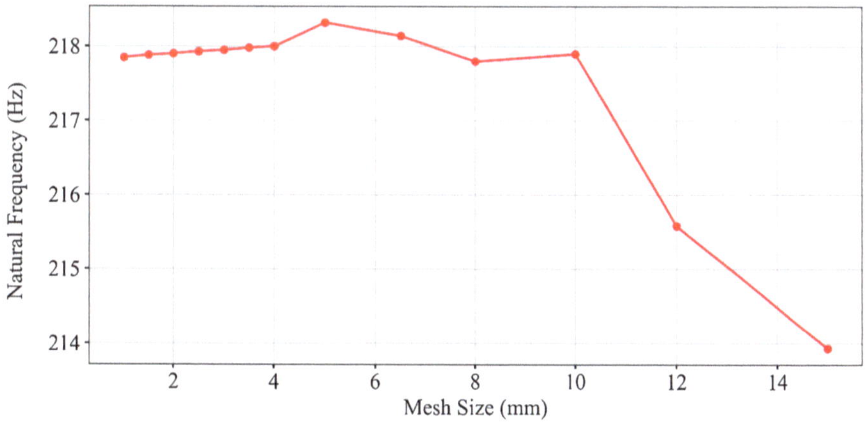

Fig. 10 Control Pipe - Mesh Convergence (Mode 2)

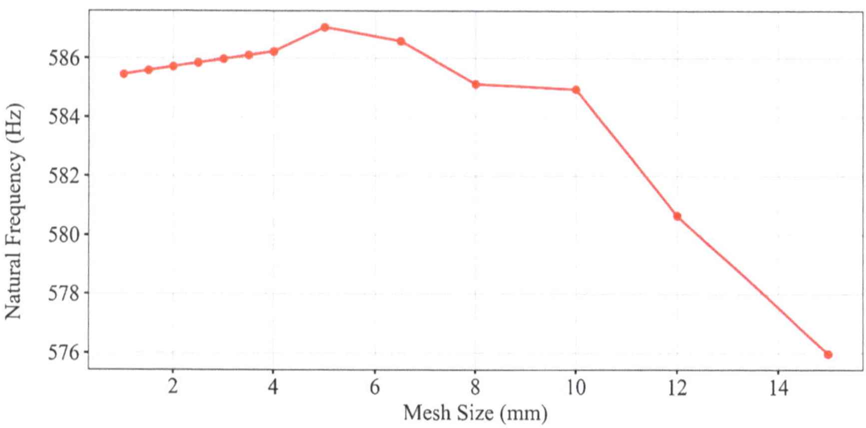

Fig. 11 Control pipe-mesh convergence (Mode 3)

2.1.4 Modal Analysis and Frequency Extraction

To understand the vibrational behavior of the pipe model and to generate the frequencies for the defect localization, a modal analysis was done for each pipe. The aim was to get the structural natural frequency, which is sensitive enough for local changes caused by defects as demonstrated in the literature review.

All frequencies were simulated below 3500 Hz. This frequency range was chosen because in a real-life experiment, practical vibration sensors can capture within that range as shown in Lengvarský et al. [40] study.

No custom changes have been made to the solver settings. The system was solved using ANSYS's Block Lanczos eigenvalue extraction method by default. No damping properties or material loss was defined; it the simulation was assumed to

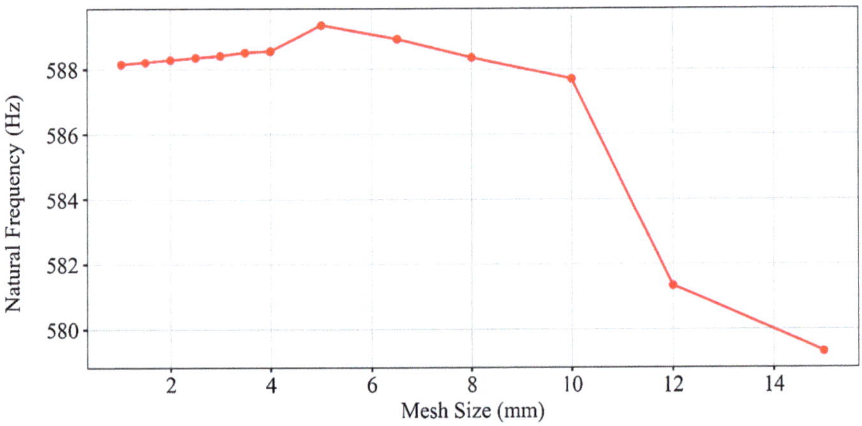

Fig. 12 Control pipe-mesh convergence (Mode 4)

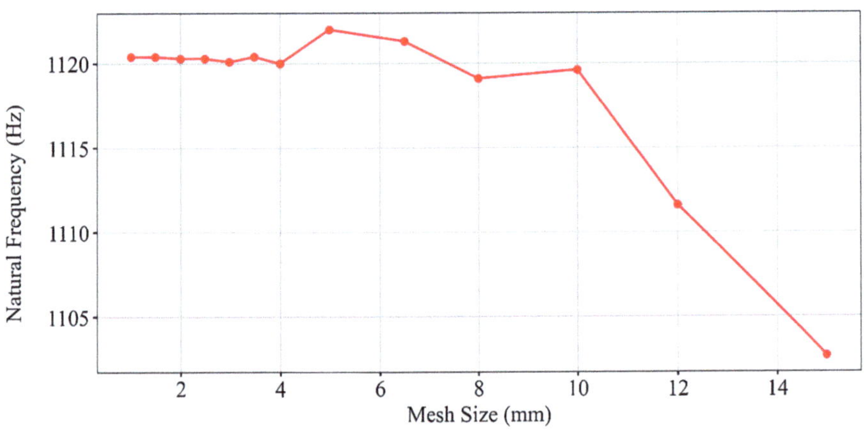

Fig. 13 Control pipe-mesh convergence (Mode 5)

be an undamped, linear elastic system, which is standard practice in structural modal evaluations [41].

Governing Equations of Modal Analysis

The dynamic equation of motion for a linear system by Craig and Kurdila [41] is given by:

$$[\boldsymbol{M}]\ddot{\boldsymbol{u}}(t) + [\boldsymbol{C}]\dot{\boldsymbol{u}}(t) + [\boldsymbol{K}]\boldsymbol{u}(t) = \boldsymbol{0}. \quad (7)$$

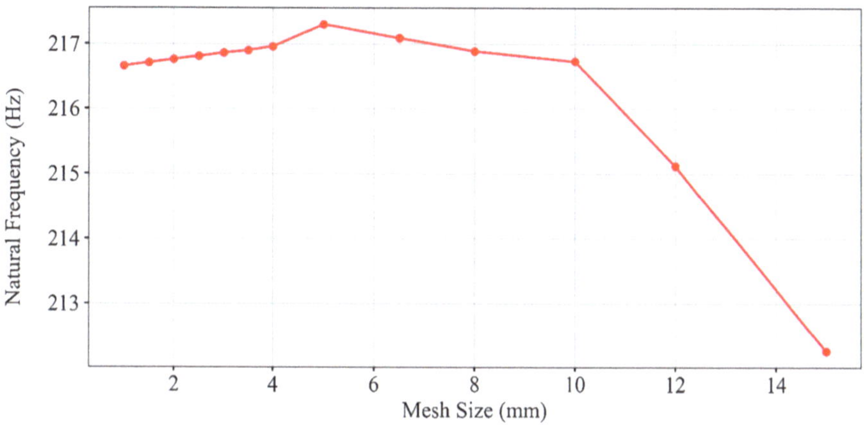

Fig. 14 Defected pipe-mesh convergence (Mode 1)

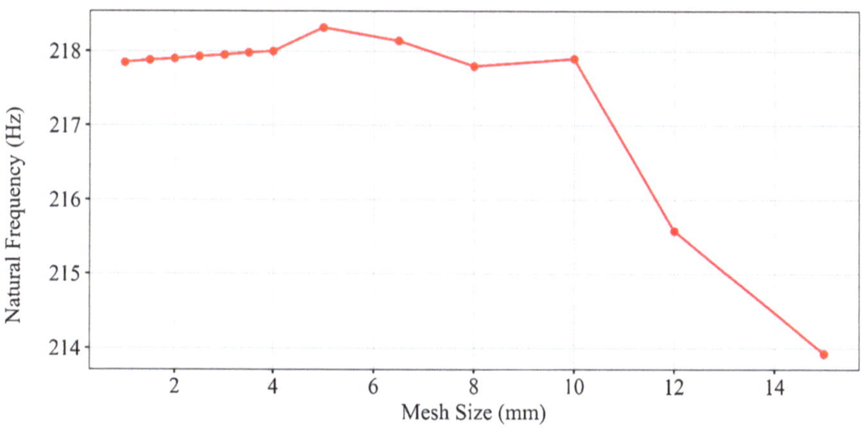

Fig. 15 Defected pipe-mesh convergence (Mode 2)

In modal analysis, the focus is on free, undamped vibration, where the damping matrix $[C]$ is assumed to be zero and no external force is applied. Simplifying the equation as described by Craig and Kurdila [41] to:

$$[\boldsymbol{M}]\ddot{\boldsymbol{u}}(\boldsymbol{t}) + [\boldsymbol{K}]\boldsymbol{u}(\boldsymbol{t}) = \mathbf{0}. \tag{8}$$

Assuming a harmonic solution in the form of $u(t) = \varnothing\, e^{iwt}$, and substituting it into the equation:

$$[\boldsymbol{K}]\varnothing = \boldsymbol{\omega}^{2}[\boldsymbol{M}]\varnothing. \tag{9}$$

This leads to a standard generalized eigenvalue problem [60]:

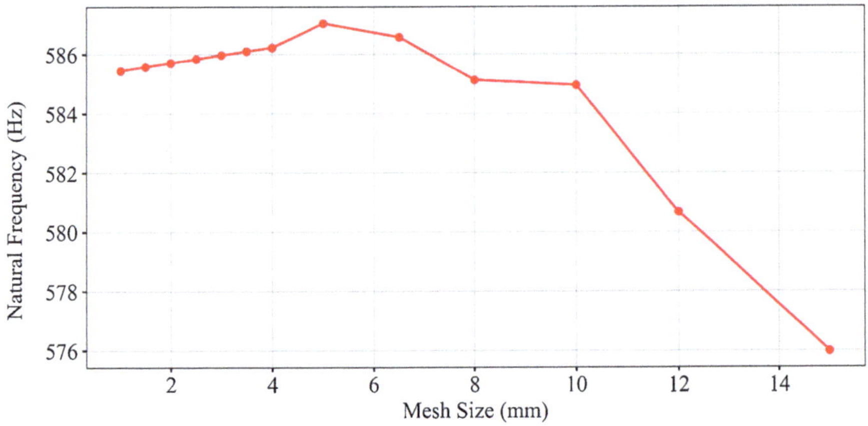

Fig. 16 Defected pipe-mesh convergence (Mode 3)

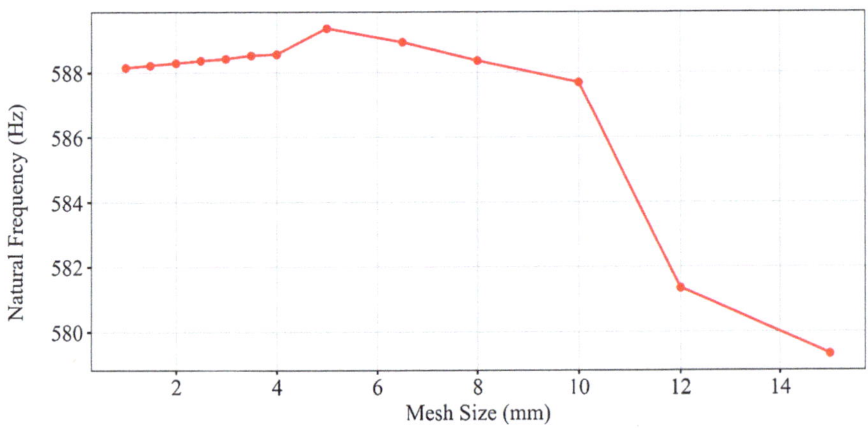

Fig. 17 Defected pipe-mesh convergence (Mode 4)

$$[\boldsymbol{K}]\boldsymbol{u} = \boldsymbol{\lambda}[\boldsymbol{M}]\boldsymbol{u}. \tag{10}$$

$$f = \frac{\sqrt{\lambda}}{2\pi}. \tag{11}$$

The parameters used in Eqns. (7), (8), (9), (10), and (11) are defined as the following:

- $[M]$ is the mass matrix
- $[C]$ is the damping matrix
- $[K]$ is the stiffness matrix
- $u(t)$ is the displacement vector
- $\dot{u}(t)$ is the velocity vector (first derivative of displacement)

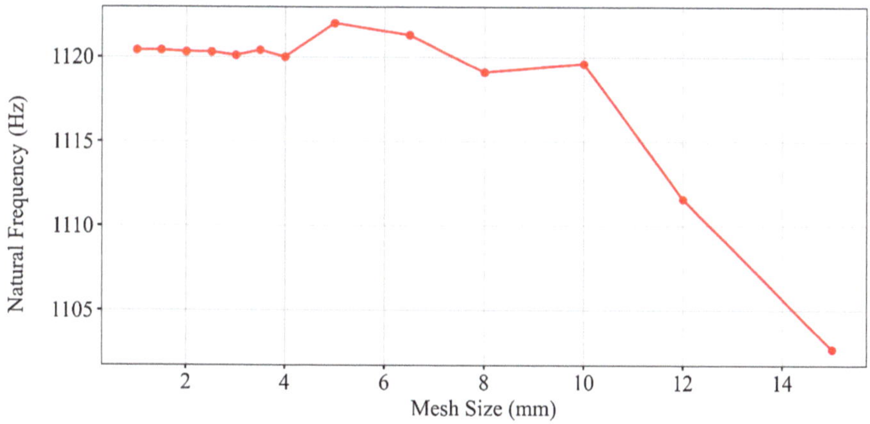

Fig. 18 Defected pipe-mesh convergence (Mode 5)

- $\ddot{u}(t)$ is the acceleration vector (second derivative of displacement)
- $\varnothing$ is the mode shape (eigenvector)
- ω is the circular natural frequency (rad/s)
- f is the natural frequency (Hz)
- i is the imaginary unit, $i = \sqrt{-1}$

The eigenvalue problem was solved by using Ansys's default Block Lanczos method, a numerical algorithm method designed for large, sparse systems. This method transforms the generalized eigenproblem into a smaller tridiagonal form using orthogonal projections and block iterations. This allows us to find the lowest number of eigenvalues. [42, 43].

Modal Setup and Frequency Extraction

The first 15 natural frequencies were obtained using ANSYS Mechanical. This number was selected for several reasons:

1. Frequency Range limitation: All 15 modes were within the range of 0–5 KHz, which assured that, it would be within the aim of the research.
2. Data Sufficiency for Neural Network Input: The 15 frequency values are a good number of inputs for the network to be trained on, providing 120 samples to each position.

The same modal setup was done for all the pipe models, and frequencies were extracted manually from the ANSYS solution results and copied to an Excel sheet to be processed later.

However, the first two natural frequencies of the control pipe were identical to the next pair as shown in Table 8. This was due to the symmetrical model and indicates

Table 8 Natural frequency of control and defect pipe with signature fault

Number of mode	Control pipe frequency (Hz)	Defected pipe frequency (Hz)	Signature fault
1	217.9	217.09	−0.81
2	217.9	218.14	0.24
3	588.41	586.59	−1.82
4	588.42	588.96	0.54
5	1124.1	1121.3	−2.8
6	1124.1	1124.9	0.8
7	1646	1643.8	−2.2
8	1802.3	1798.8	−3.5
9	1802.3	1802.9	0.6
10	2601.8	2597.8	−4
11	2601.8	2601.6	−0.2
12	2654.1	2652.5	−1.6
13	3292	3287.5	−4.5
14	3502.3	3499.3	−3
15	3502.3	3500.8	−1.5

mode duplication due to axis symmetry. In a real-life experiment this symmetry would be impossible, since slight asymmetric are present pipes during manufacturing as proved in Lengvarský et al. [40] study.

To prepare the input for the neural network, the extracted natural frequency of each defective model was subtracted from the control model to give us the fault signature vectors as given in Table 8. This will be discussed in the next section.

2.1.5 Neural Network Training and Defect Prediction

Problem Formulation and Network Objective

The original objective was to predict both the 1D longitudinal and the axial-circumferential position of defects using only differences in natural frequencies; however, this was later narrowed to estimating just the 1D longitudinal position (in millimeters). This is supported by Mohammadzaheri et al. [44], where he tested locating a fault along a single dimension in a complex engine component using resonance frequencies below 5 KHz in a fuzzy interference. The results indicated that these frequency changes can be used to locate the fault and setting them up as inputs for data-driven models.

The ANN was formulated as a regression model rather than a classifier, since the output is intended to be a continuous position estimation ranging from the value of 39 mm to 497 mm.

Neural Network Output Design

The ANN was originally designed to produce two continuous output value longitudinal position and circumferential axis using all 15 modal frequency shifts as inputs. However, the network struggled to see the difference between mirror locations, especially since larger defects sizes produced very small frequency shifts pattern across different circumferential positions. A second approach was made; this was done by splitting defect prediction into two separate networks each with 15 inputs and 1 output: one predicted longitudinal position and the other circumferential axes. After simplifying each regression task, neither the configuration estimates the circumferential axes accurately, so the final implementation was to use only the single output network in predicting the longitudinal defect position.

In addition, only half of the pipe length for the defective localization was done to maintain a clean dataset and avoid redundant mirrored cases, so only the defects were placed in the first half of the pipe.

Input Features and Data Preprocessing

The input feature vector of each sample was 15 values, calculated by taking the difference between the defective pipe and the control pipe's natural frequency. Providing a 120 × 15 matrix.

$$ {}_kS_i = f_{defect,i} - f_{control,i}, \quad \boldsymbol{i = 1, 2, \ldots 15}. \tag{12} $$

Each sample was given two labels: the position of a defect in the pipe's length (39 to 497 mm) and an index for circumferential position (integer 1–8 for each of eight spots around the pipe).

Normalization was done by choosing z-score normalization on the input data, because this feature standardizes each feature to have zero mean and unit variance. This helps gradient-based learning algorithms to be more reliable and converge faster [45].

The output values were left unnormalized and kept in millimeters. This avoids any post-processing since the physical meaning was reserved. Furthermore, since the output was a single continuous value, it remained within a numerically manageable range for the network to learn. In addition, the output scale is small, which wouldn't affect the coverage speed or introduce numerical instability [45].

For models that attempted to predict the circumferential axis, the axis label (1–8) was treated as a separate continuous target; however, as explained in Sect. 3, geometric symmetry issues ultimately led to using only the longitudinal position label.

Neural Network Architectures and Configuration

Since the relationship between the small shifts in modal frequencies and defect position is highly nonlinear. ANN was selected as proven to be a universal approximation property by Hornik et al. [46] and has shown their success in locating vibration-based fault localization in pipe and other structures [47, 48]. Their ability to learn complex, nonlinear mappings directly from frequency-domain features to continuous position estimates makes them a good option for this task.

Three different network architectures were implemented:

1. Feedforward Neural Network (`feedforwardnet`)
2. Cascade-Forward Neural Network (`cascadeforwardnet`)
3. Curve Fitting Network (`fitnet`)

Each of the network architectures shared the same structural layout:

- 15 input neurons (one for each frequency difference)
- Two hidden layers with 10 neurons each
- 1 output neuron for the predicted defect position

The activation functions were:

- Hidden layers: tansig (hyperbolic tangent sigmoid)
- Output layer: purelin (linear), suitable for continuous regression outputs

The Feedforward neural network has been widely used in structural and frequency-based modeling, as demonstrated by Pettersson et al. [49]. The author applied the feedforward neural network to predict the average sound pressure from the frequency domain data. This serves as a strong foundation for studying modal frequencies and defect positions.

The idea to use the cascade-forward neural network (CFNN) in this study came from the research done by Kourehli [50] on identifying damage in structures using frequency-based features during short-term dynamic loading. For this reason, CFNN was used in this work to predict the locations of defects in pipes using modal frequency results. Having more connections between the layers in CFNN than feedforward networks can help with how the model is trained and makes it suitable for the task.

Fitnet was chosen due to it specialization in regression tasks. This model shares the same network structure as feedforward, however its tuned specifically for regression in MATLAB. This is done through the internal configurations that can lead to slightly better weight initialization, scaling behavior, and training stability, even when all visible settings are the same. This made it more effective for predicting continuous defect positions. The choice was inspired by Saha and Giurgiutiu [48], where the authors were successful in using MATLAB fitnet tool to model structural damage based on frequency data [48].

Equation (13) was used to initially estimate the (h) number of neurons in the hidden layers, where (s) indicates the number of input features, this equation can be used as a starting point according to Mohammadzaheri et al. [51]. Having 15 inputs, the first layer had a size of 31 neurons. This served as a basis for exploring different network configurations. Since no single structure guarantees optimal performance, multiple architectures were tested in a trial-and-error approach. During this process, it was found that introducing a second hidden layer improved training stability and generalization. Research indicates that deeper architecture often learn complex more pattern efficiently leading to better results [52]. Stathakis [53] supports this approach, hence it's a valid and commonly adopted method for selecting hidden layers and neurons.

$$\boldsymbol{h} = \mathbf{2}\boldsymbol{s} + \mathbf{1}. \tag{13}$$

The hidden layer activation function was chosen because of its regular usage in similar regression tasks. [61] demonstrated that tansig activision function, results in a rapid training and accurate predictions in feedforward backpropagation networks. Specifically, tansig is defined as

$$\boldsymbol{tansig}(\boldsymbol{n}) = \frac{\mathbf{2}}{\mathbf{1} + \boldsymbol{e}^{-\mathbf{2}\boldsymbol{n}}} - \mathbf{1}. \tag{14}$$

Where it compresses the inputs *(n)* into the interval (−1,1). This makes it particularly suitable for handling nonlinear regression of vibration-related data. This is also shown by Saha and Giurgiutiu [48] study, where the function was used for the hidden layers for creating their neural network. It is evident from these studies that tansig is an appropriate choice for handling nonlinear regression of data related to vibrations or dynamic systems.

Purelin activation function was used in the output layer as inspired by Xie et al. [54] proved that it helps the network avoid the problems with saturation and limited range of sigmoid activations and the predictions can match real, continuous measurements without rescaling. In mathematical form, purelin is simply:

$$\boldsymbol{purelin}(\boldsymbol{n}) = \boldsymbol{n}. \tag{15}$$

so that the network's final output is identical to its input n. The model outputs the defect locations in millimeters and this helps both with understanding and with the model converging steadily during optimization.

Mathematical Model of the ANN

The ANN developed in this study is a regression model that maps a 15-dimensional input vector to output a representation of the defect locations in millimeters, The mathematical structure of the network with two hidden layers.

Feedforward Neural Network Model

The standard feedforward neural network (FFNN) processes input through successive fully connected layers, each followed by a nonlinear activation function. The forward propagation of the network is given by:

$$\begin{aligned} z_1 &= W_1 x + b_1 \\ a_1 &= \sigma_1(z_1) \\ z_2 &= W_2 a_1 + b_2 \\ a_2 &= \sigma_1(z_2) \\ \hat{y} &= \sigma_2(W_3 a_2 + b_3) \end{aligned} \tag{16}$$

This can also be written compactly as:

$$\hat{\boldsymbol{y}} = \boldsymbol{f}(\boldsymbol{x}) = \boldsymbol{W_3} \cdot \boldsymbol{\sigma_2}(\boldsymbol{W_2} \cdot \boldsymbol{\sigma_1}(\boldsymbol{W_1} \cdot \boldsymbol{x} + \boldsymbol{b_1}) + \boldsymbol{b_2}) + \boldsymbol{b_3}. \tag{17}$$

where

- $x \in R^{15}$ is the input vector (the fault signature)
- $\hat{y} \in R$ is the predicted defect position
- $W_1,\ W_2, W_3$ are the weight matrices for each layer
- $b_1,\ b_2, b_3$ are the bias vectors
- σ_1 is the activation function used by hidden layer (tansig)
- σ_2 is the activation function for the output layer (pureline)

Levenberg–Marquardt (LM) backpropagation was applied because it combines the fast speed of Gauss–Newton and stability from gradient descent which makes it suitable for use in small- to medium-sized networks [55]. LM's adaptive damping avoids local minima and accelerates training, often requiring fewer epochs than standard methods [56]. In a similar manner, Kessai et al. [47] relied on LM-trained ANNs to spot cracks in drilling pipes by analyzing their vibration data, showing the method to be useful in such tasks. The parameter update rule is:

$$\Delta\boldsymbol{\vartheta} = -\left(\boldsymbol{J}^T\boldsymbol{J} + \boldsymbol{\mu}\boldsymbol{I}\right)^{-1}\boldsymbol{J}^T\boldsymbol{e}. \tag{18}$$

where:

- e is the error vector
- J is the Jacobian of the error w.r.t. parameters
- μ is the damping parameter
- I is the identity parameter

This method is particularly effective for small-to-medium networks and was used across all ANN models. The total number of trainable parameters in each network architecture is 281, this includes all the weights and biases.

The weights and biases of the neural networks were initialized using the Nguyen–Widrow method, which enhances training speed and convergence in feedforward networks with sigmoid-type activations like tansig. This method distributes weights to keep neuron activations within their sensitive range. It is commonly adopted in MATLAB's feedforward net and was applied across all networks in this study. A similar approach was used by Saha and Giurgiutiu [48], for detecting structural damage using ANN models, demonstrating its effectiveness in vibration-based fault diagnosis.

Cascade-Forward Neural Network Model

The **cascade-forward neural network (CFNN)** is structurally like the feedforward network but includes additional shortcut connections from the input layer to each hidden and output layer. These direct connections allow the raw input to reach deeper layers, improving convergence speed and learning efficiency, especially in small or complex datasets.

The propagation through the cascade-forward network is described as:

$$\hat{y} = \boldsymbol{W}_3 \cdot \sigma_2(\boldsymbol{W}_2 \cdot \sigma_1(\boldsymbol{W}_1 \cdot \boldsymbol{x} + \boldsymbol{C}_1 \cdot \boldsymbol{x} + \boldsymbol{b}_1) + \boldsymbol{C}_2 \cdot \boldsymbol{x} + \boldsymbol{b}_2) + \boldsymbol{C}_3 \cdot \boldsymbol{x} + \boldsymbol{b}_3. \tag{19}$$

where:

- C_1, C_2, C_3 are weight matrices for the shortcut (cascade) connections
- All other symbols follow the same notation as in Eqn (17)

The CFNN uses the same **tansig** and **purelin** activation functions, and the same LM backpropagation training method. The additional cascade connections increase the model's ability to capture complex mappings between input and output, potentially reducing the number of required epochs during training.

Fitnet Architecture for Function Approximation

The fitnet model is designed for regression and function fitting. In this study, it was configured with two hidden layers, consistent with the feedforward and cascade-forward models to ensure fair comparison.

Its mathematical model and activation flow are identical to the feedforward network (Eqns. (16), (17), (18)), using tansig in the hidden layers and purelin in the output. Training was performed using the Levenberg–Marquardt algorithm. Fitnet's structure and performance make it well-suited for capturing nonlinear patterns in vibration-based defect prediction.

Training Process

The 70:15:15 split was done and is a common practice in machine learning, hence the reason it was selected [57]. Stratified technique was implemented to ensure that all 15 discrete defect position (39 mm to 497.03 mm), were divided equally in the training, validation and test sets. A random split could result in an over or under sample in certain locations, causing the network to see some locations or only during evaluation, which would bias its performance metrics and impair generalization. By organizing samples into groups based on their true axial index and then allocating a consistent 70/15/15 split within each group.

The method preserves the original output distribution and yields more reliable performance estimates, ensuring that the network learns from and is evaluated on examples covering the entire length of the pipe [58, 62].

During the training, no custom performance goal or validation-failure limit was set. MATLAB automatically stooped training when the performance function gradient dropped below its default minimum gradient threshold 1×10^{-7}, meaning further weight updates were negligible. During the training the epochs reached 14 for both feedforward and fitnet, and at epoch 8 for cascade-Forward. Relying on minimum-gradient criterion is suitable for moderate networks, as if any additional training was done, it could result in overfitting [56]. This method also helps in conserving computational power by stopping the training once genuine threshold is reached, which matches best practices for Levenberg–Marquardt–based ANN training [55].

Several studies in vibration-based damage localization (e.g., [47, 48]) have similarly relied on MATLAB's default LM implementation by stooping it when the training-error gradient falls below 1×10^{-7} demonstrating that this stopping criterion is both common and effective for moderate-sized networks.

Artificial Neural Network Structures

The configuration used for defect position prediction, showing 15 normalized inputs, two hidden layers using tansig activation, and one linear output neuron for all three networks.

Figure 19 shows the neural network architecture: (a) Feedforward neural network, (b) Cascade-forward neural network, (c) Fitnet neural network.

2.1.6 Linear Regression Modeling

Following the development of the ANN models, a linear regression approach was implemented to estimate the axial position of pipeline defects using the same dataset. The goal was to evaluate how a simpler, interpretable model would perform on the same frequency-based features used for ANN training. Two different data splits were applied: Dataset A with 70% training and 30% testing, and Dataset B with 80% training and 20% testing, to examine the influence of data partitioning on model performance.

The model was implemented in MATLAB, which provides robust support for regression, matrix computation, and statistical evaluation. The linear regression

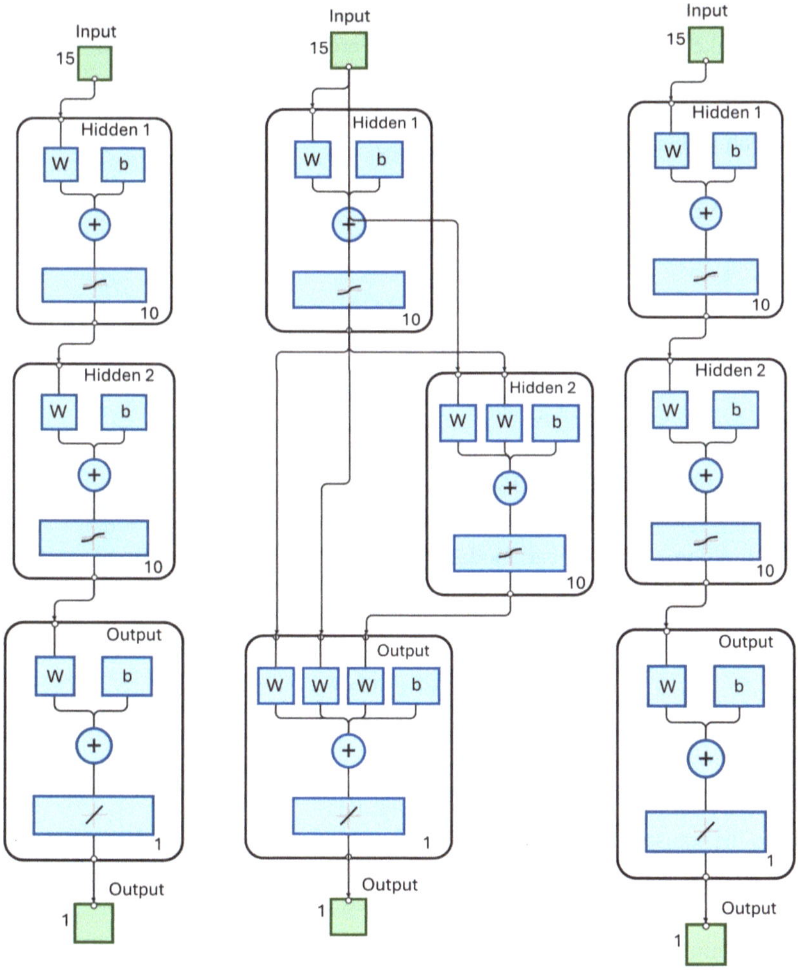

Fig. 19 Neural network architectures

setup shared the same input features (15 normalized frequency shifts), output target (defect location in mm), data preprocessing methods, and splitting techniques as used in the ANN models. This consistency ensured a fair comparison across modeling approaches.

Linear regression was selected due to its effectiveness in mapping predictable numerical relationships and its interpretability in engineering contexts. The fitlm() function in MATLAB was used to create the model, offering a statistically reliable and versatile interface. Unlike basic linear regression tools, fitlm() provides insight into the model's behavior by outputting regression coefficients, confidence intervals, and statistical diagnostics [59]. Additionally, it generates residual and leverage plots, useful for detecting data irregularities and validating model assumptions [63]. These features make it well-suited for academic research.

Mathematical Foundation of Training

The linear regression model predicts the defect location $\widehat{y}$ as a weighted sum of the 15 input features (x_1 to x_{15}), following the general linear equation:

$$\widehat{y} = \boldsymbol{\beta}_0 + \boldsymbol{\beta}_1 \boldsymbol{x}_1 + \boldsymbol{\beta}_2 \boldsymbol{x}_2 + \ldots + \boldsymbol{\beta}_{15} \boldsymbol{x}_{15}. \quad (20)$$

where

- $\hat{y}$ is the predicted defect location
- x_1 to x_{15} is the normalized frequency signature inputs (S_1 to S_{15})
- β_0 is the intercept (bias term)
- β_1 to β_{15} are the coefficients (learned weights from training)

The training of the linear model involves computing the optimal coefficient vector β that minimizes the squared difference between predicted and actual defect locations. This is achieved using Normal Equation:

$$\boldsymbol{\beta} = \left(\boldsymbol{X}^T \boldsymbol{X}\right)^{-1} \boldsymbol{X}^T \boldsymbol{y}. \quad (21)$$

where

- X is the matrix of input features (training data)
- X^T is the transpose of X
- y is the vector of actual output values (target locations)
- β is the vector of learned coefficients (including intercept)

This formula allows MATLAB's regression system (or other numerical tools) to find the linear hyperplane that fits the data best by least squares.

2.1.7 Evaluation and External Testing

The model's performance was evaluated using the following metrics:

- Mean absolute error (MAE)
- Root mean squared error (RMSE)
- Coefficient of determination (R^2)
- Mean absolute percentage error (MAPE)

In addition, nine extra models were created at positions of 54 mm, 98 mm, 123 mm, 137 mm, 181 mm, 212 mm, 289 mm, 315 mm, and 371 mm, which were set aside as an external test set. These models were not used during the training or the validation and were intended to determine the effectiveness of the models. The same test is done on the linear regression to see how well it generalizes.

2.1.8 Methodology Summary

This study used ANSYS-based simulations to diagnose pipeline defects through vibration analysis. A total of 121–130 steel pipe models were created, each with a single hole at varying longitudinal and circumferential positions, along with a defect-free control pipe. Modal analysis was performed to extract the first 15 natural frequencies from each model, all under 5 kHz to match standard vibration sensor capabilities. Input features were calculated as the frequency differences between each defective pipe and the control pipe, forming 15-dimensional vectors. The first modeling approach applied three artificial neural network (ANN) architectures—feedforward, cascade-forward, and fitnet—each with two hidden layers of 10 neurons, trained using the Levenberg–Marquardt algorithm. The hidden layers used the tansig activation function, while the output layer used purelin for continuous regression. Performance was evaluated using MAE, RMSE, MAPE, and R^2, including tests on nine unseen pipe configurations. To complement the ANN results, a linear regression model was developed in MATLAB using the same input-output structure and evaluated with the same metrics and data splits (70/30 and 80/20). This dual-method approach enabled a robust comparison of predictive accuracy and generalization, supporting the integration of AI in vibration-based structural health monitoring.

3 Section 3

3.1 Results and Discussion

3.1.1 Overview

This section presents the results of both the artificial neural network (ANN) models and the linear regression model, each trained to predict longitudinal and axial circumferential defect positions in pipelines using low-frequency modal analysis data. The ANN evaluation includes three architectures—feedforward, cascade-forward, and fitnet—trained using a fixed 70/15/15 data split. In contrast, the linear regression model was trained and tested on two separate data splits: 70% training with 30% testing, and 80% training with 20% testing. This was done to observe how different data partitions affect the linear model's performance.

The section is organized into four main parts. The first three parts focus on the ANN models. For each architecture, the results are divided into three subsections: internal test set performance, generalization to nine unseen longitudinal defect cases, and performance in detecting axial circumferential defect positions. The fourth part presents the results of the linear regression model, also evaluated across longitudinal and axial defect predictions using both data splits.

All models are assessed using standard regression metrics including mean absolute error (MAE), root mean square error (RMSE), mean absolute percentage error (MAPE), and the coefficient of determination (R^2). These metrics help quantify prediction accuracy and model consistency. Throughout the section, comparative analysis is provided to determine which modeling approach yields the best balance between predictive accuracy and generalization. Visual tools such as performance tables, residual plots, and regression figures are used to support the findings and enhance clarity.

3.1.2 Feedforward Neural Network

This section will present the prediction performance of the feedforward neural network based on internal test and a separate test consisting of nine unseen samples. The results are evaluated based on the standard regression metrics and visualized through different plots.

In terms of choosing the optimal number of neurons and layers, feedforward neural network was used as a benchmark. Since it's the simplest form of regression model: inputs flow straight through one or more fully connected layers to the output, without any extra feedback loops or scaling layers. This gives an advantage to test several different layer sizes and neuron counts in a short time.

Below is a summary of the feedforward configurations tried, alongside their validation errors:

As given in Table 9, a series of tests were made to check the best architecture for the network. The table summarizes the different hidden layer architecture used during the testing. The results showed that the two-layer [10 10] network gave the lowest MSE (1.08 mm^2) and MAE (0.85 mm). It gave a balance capacity to learn the frequency shifts and connect it to the defection position, without becoming too large or overfitting. The same setup was used for the rest of the networks.

Table 9 Neural network performance by architecture

Hidden-layer Architecture	MSE (mm^2)	MAE (mm)
[31]	1.50	1.05
[25]	1.40	0.98
[20]	1.30	0.92
[15]	1.20	0.88
[10 10]	1.08	0.85
[20 10]	1.15	0.90
[30 15]	1.12	0.89
[5 5]	1.35	1.00

Internal Test Set Performance

Table 10 provides the predicted versus the actual defect position for 18 external test samples. As observed, the model achieved a strong prediction accuracy across the different positions. Slightly high errors can be seen around 4.4% at the pipe edge (39 mm).

Figure 20 presents the regression plots for training, validation, and test sets. The regression lines closely follow the $y = x$ line up in all three splits, with the R^2 values, exceeding 0.999. This shows that the network is an excellent fit and achieved a minimal variance across the datasets.

- X-axis: “Target (mm)”
- Y-axis: “Output (mm)”

Figures 21 and 22 show the predicted vs actual values and with their residuals. As shown, the residuals are randomly distributed and centered around zero, indicating no apparent bias in predictions.

The error histogram shown in Fig. 23 demonstrates that most of the residuals fall within a narrow range and are symmetrically distributed. This confirms that the network is consistent and stable in its prediction.

Figure 23 summarizes the training behavior of the network. The gradient steadily decreased with few validation checks, and the training process converged smoothly, confirming a well-behaved learning curve without signs of overfitting.

Table 10 Feedforward internal test set performance

Axes label	True value (mm)	Predicted value (mm)	Absolute error (mm)	Relative error (%)
1	39.000	40.694	1.6942	4.3442
7	39.000	37.265	1.7347	4.4478
4	71.760	71.663	0.09651	0.13449
5	71.760	71.494	0.26614	0.37088
8	71.760	71.321	0.43940	0.61232
2	71.760	71.605	0.15458	0.21541
1	104.39	104.39	0.000	0.000
4	104.39	106.47	2.0829	1.9953
6	104.39	102.34	2.0532	1.9669
8	169.83	170.10	0.26763	0.15759
3	267.99	267.69	0.30074	0.11222
1	300.71	301.17	0.45955	0.15282
7	300.71	300.59	0.12245	0.04072
2	333.43	333.43	0.000	0.000
4	366.15	365.60	0.55334	0.15112
6	398.87	398.61	0.26263	0.06584
1	497.03	496.77	0.25776	0.05186
3	497.03	496.80	0.23043	0.04636

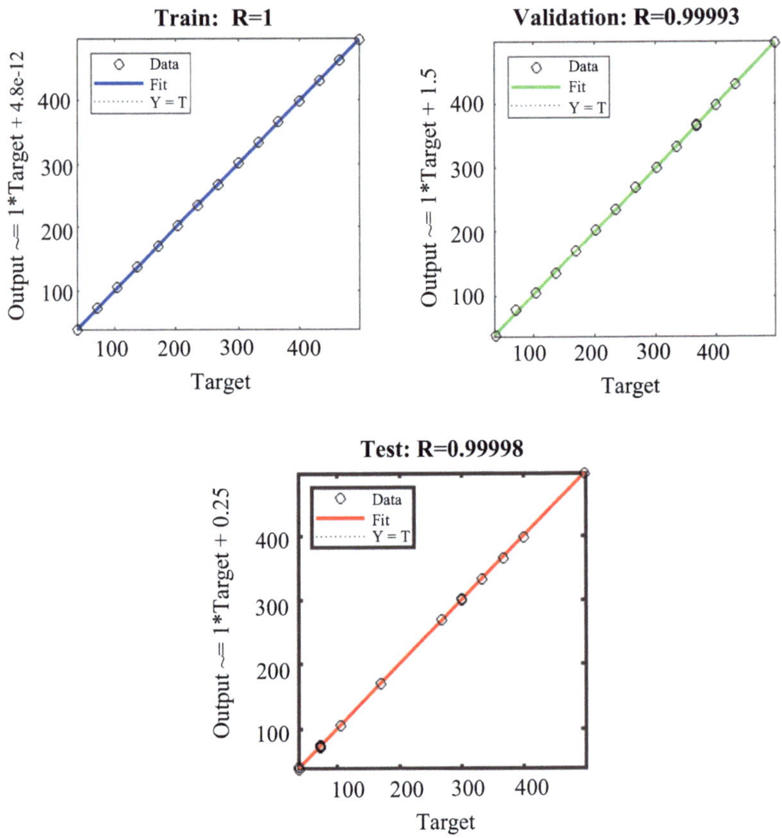

Fig. 20 Feedforward regression plots

Figure 24 shows the training plot summary for the network. The gradient falling to ~1.16×10^{-9} by epoch 14 indicates the network's error surface has effectively flattened, signaling convergence.

In terms of numerical performance, this is what the feedforward network achieved (Table 11)

The close agreement in performance across the training, validation, and test sets—each exhibiting low error values and high R^2—demonstrates that the model generalized well and is not overfitting

External Test Set Performance

Table 12 compares the actual and forecasted defect positions in the nine external test samples. The model demonstrated a strong generalization, as most of its errors

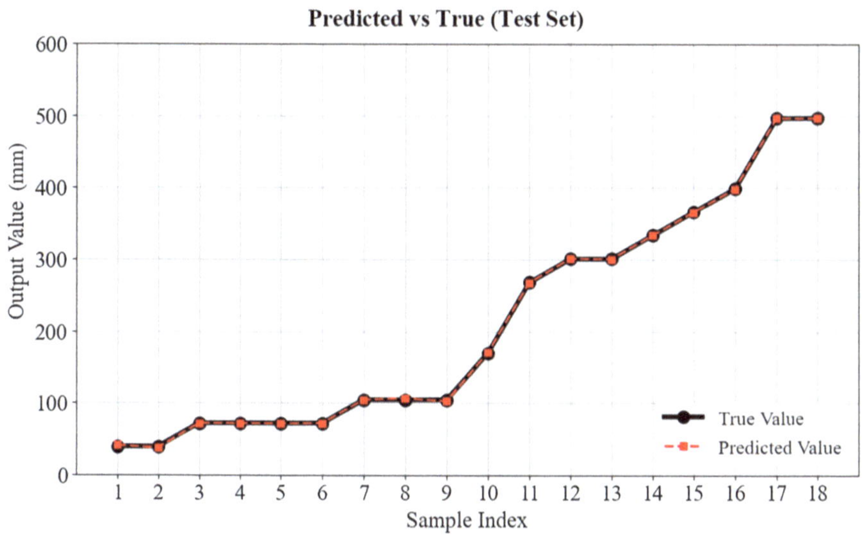

Fig. 21 Feedforward neural network internal test set performance

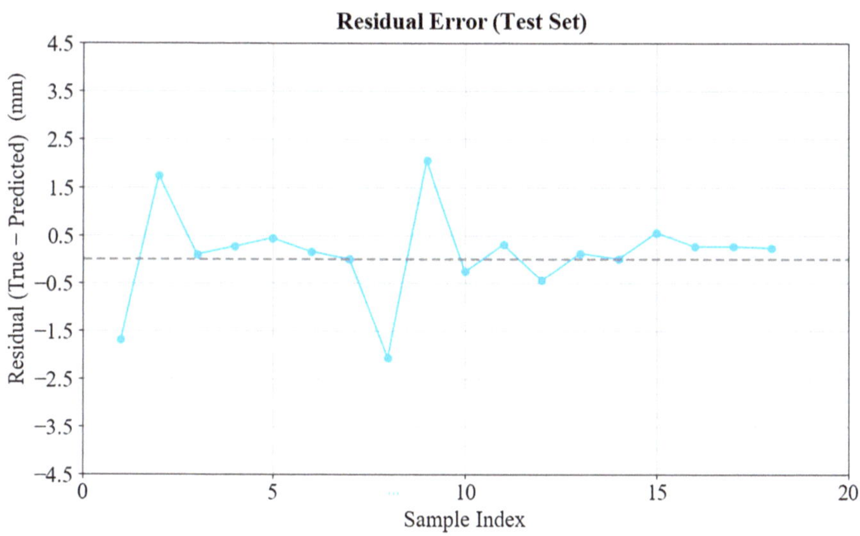

Fig. 22 Feedforward neural network residual errors

stayed low. The largest absolute error was 14.22 mm at the 212 mm point, which gives a relative error of 6.89%.

Figure 25 shows the predicted versus true values for these unseen samples. Most predictions fall close to the identity line, but we can see a deviation at the center of the graph like in samples 4, 6, and 8.

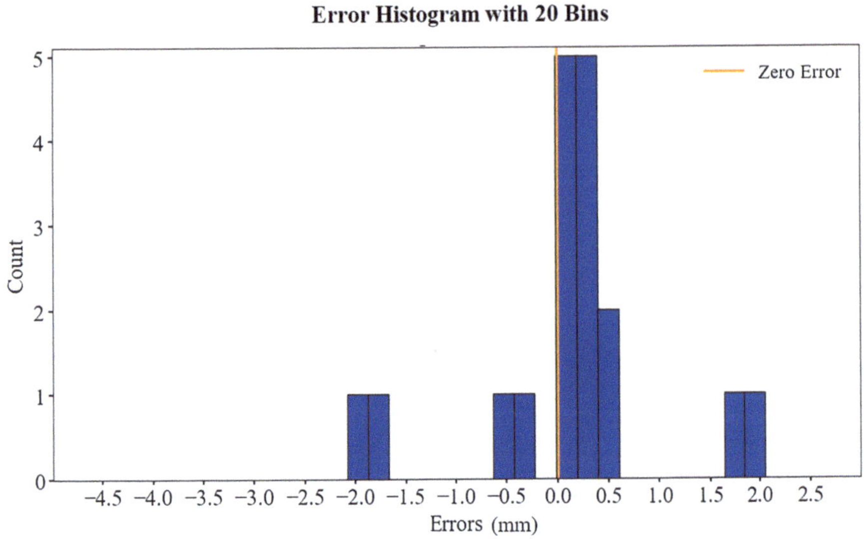

Fig. 23 Feedforward neural network error histogram

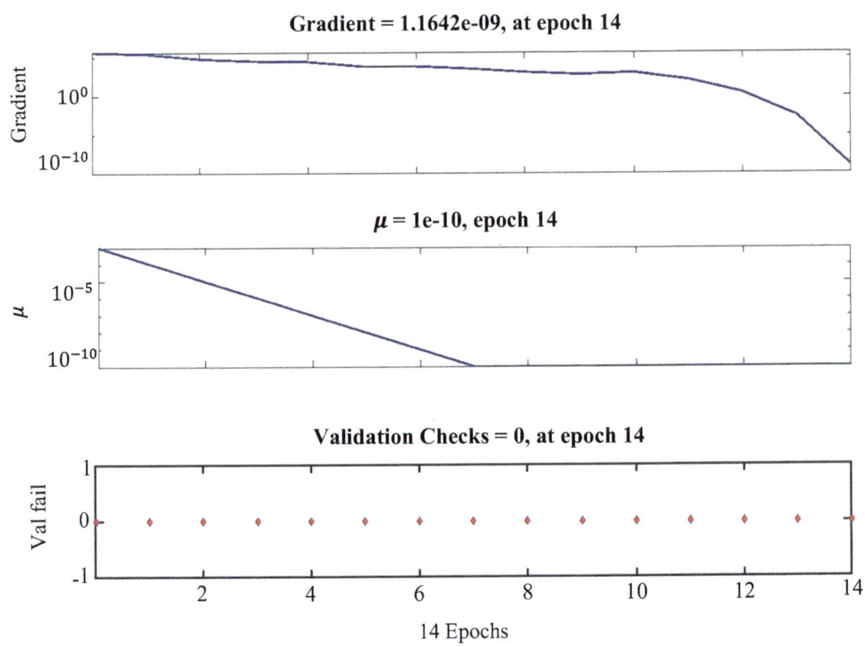

Fig. 24 Feedforward neural network training state plot

Table 11 Feedforward neural network numerical performance

Performance metrics	Training set	Validation set	Test set
RMSE	0.000 mm	1.7308 mm	1.0604 mm
MAE	0.000 mm	1.0831 mm	0.7930 mm
MAPE	0.00%	1.06%	1.00%
R^2	1.000	0.9998	1.000

Table 12 Feedforward external test set performance

True value (mm)	Predicted value (mm)	Absolute error (mm)	Relative error (%)
54.000	54.061	2.0609	3.8164
98.000	103.84	5.8435	5.9628
123.00	118.00	4.9961	4.0619
137.00	150.80	13.799	10.072
181.00	174.12	6.8812	3.8018
212.00	197.40	14.597	6.8853
289.00	294.28	5.2755	1.8254
315.00	329.22	14.219	4.5140
371.00	371.08	0.0826	0.0223

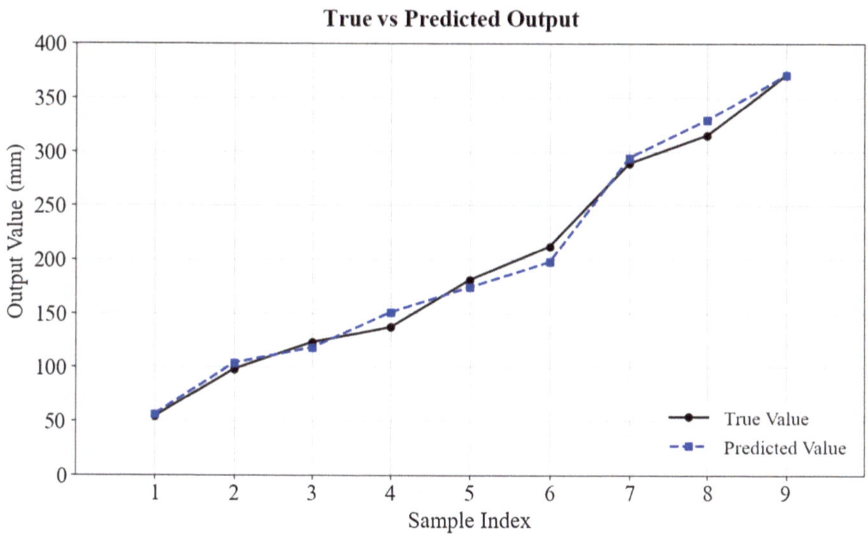

Fig. 25 Feedforward neural network external test set performance

The model achieved the following performance metrics on the external test set:

- MAE: 7.12 mm
- RMSE: 8.32 mm
- MAPE: 3.98%

Table 13 Feedforward axial detection performance

Performance metrics	Training set	Validation set	Test set
RMSE	0.5164	1.6278	1.7546
MAE	0.1333	0.5355	0.8607
MAPE	3.60%	12.63%	46.96%
R^2	0.9477	0.4724	0.5210

The model's accuracy declined on the external test set, but the results were still good enough for setting up a localization system. For the majority of the positions, the network had low relative errors (under 5%), and the larger errors were found at a limited number of areas.

Axial detection performance (Table 13)

The feedforward network fits the training data almost perfectly (RMSE $= 0.52$, MAE $= 0.13$, $R^2 \approx 0.95$), but its performance decreases on unseen data—validation errors rise to RMSE ≈ 1.63, MAE ≈ 0.54 ($R^2 \approx 0.47$), and test errors are RMSE ≈ 1.75, MAE ≈ 0.86 ($R^2 \approx 0.52$), with a MAPE of ~47%, indicating a sign of overfitting and poor generalization for precise defect axial localization

3.1.3 Cascade-Forward Neural Network

This section presents the prediction performance done by the Cascade network in estimating the defect position.

Internal Test Set Performance

Table 14 provides the internal test results of the cascade-forward network, with the majority of errors falling below 3%. Around 71.76 mm, the network error margin is slightly higher than others.

Figure 26 represents the regression plots for training, validation, and test datasets. All three of them are very close to the ideal line $y = x$, with R^2 values above 0.999. This indicates an excellent correlation and minimal variance across the data splits.

Figures 27 and 28 show the predicted vs actual values along with their residuals for the Cascade network. In Fig. 28, we can see some of the points are quite far from the zero line.

The error histogram shown in Fig. 29 has most of the errors concentrated around zero with a relatively symmetric distribution. The highest we can see tend toward -4.0, as observed most of the errors are on the negative side

Table 14 Cascade-forward internal test set performance

Axes label	True value (mm)	Predicted value (mm)	Absolute error (mm)	Relative error (%)
1	39.000	38.711	0.28914	0.74139
7	39.000	40.233	1.2328	3.1610
4	71.760	71.084	0.67566	0.94156
5	71.760	73.921	2.1631	3.0143
8	71.760	75.841	4.0810	5.6871
2	71.760	71.778	0.01803	0.02512
1	104.39	104.39	0.000	0.000
4	104.39	106.26	1.8715	1.7928
6	104.39	102.92	1.4653	1.4037
8	169.83	169.59	0.23658	0.13930
3	267.99	266.29	1.6978	0.63354
1	300.71	300.57	0.14408	0.04791
7	300.71	303.71	2.9951	0.99602
2	333.43	333.43	0.000	0.000
4	366.15	364.61	1.5352	0.41929
6	398.87	396.56	2.3147	0.58032
1	497.03	497.74	0.71257	0.14337
3	497.03	495.39	1.6391	0.32978

Figure 30 illustrates the cascade-forward network's training behavior. By epoch 8, the gradient has decreased to around 1.39×10^{-11}, suggesting that the network's error surface has essentially flattened and the learning process has converged.

Table 15 proves that the network demonstrated its capabilities in learning the fault longitudinal location patterns effectively with minimal error.

External Test Set Performance

Table 16 presents the predicted versus the true defect positions for nine samples of data. The model performed well, with small errors. The largest absolute error was approximately 10.67 mm, at a 123 mm position, with a relative error of 8.67%.

Figure 31 shows the comparison of the true and predicted positions for the external test samples. The predictions matched the true labels closely for all positions, showing that the cascade-forward network could handle new cases well.

In terms of performance, the external test set yielded the following error metrics:

- **MAE**: 7.1185 mm
- **RMSE**: 8.3162 mm
- **MAPE**: 3.98%

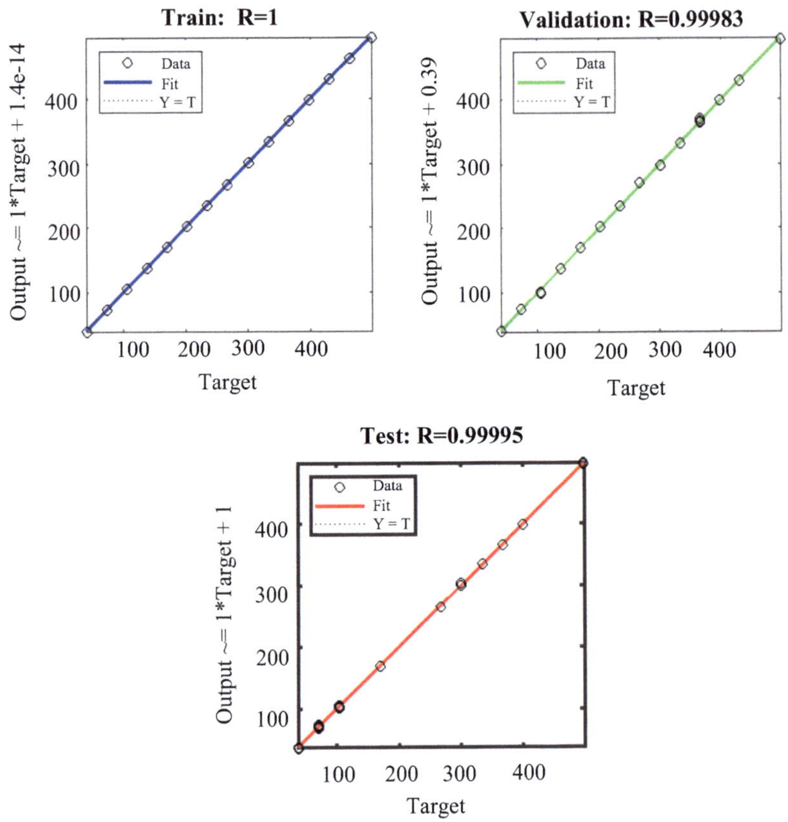

Fig. 26 Cascade-forward regression plots

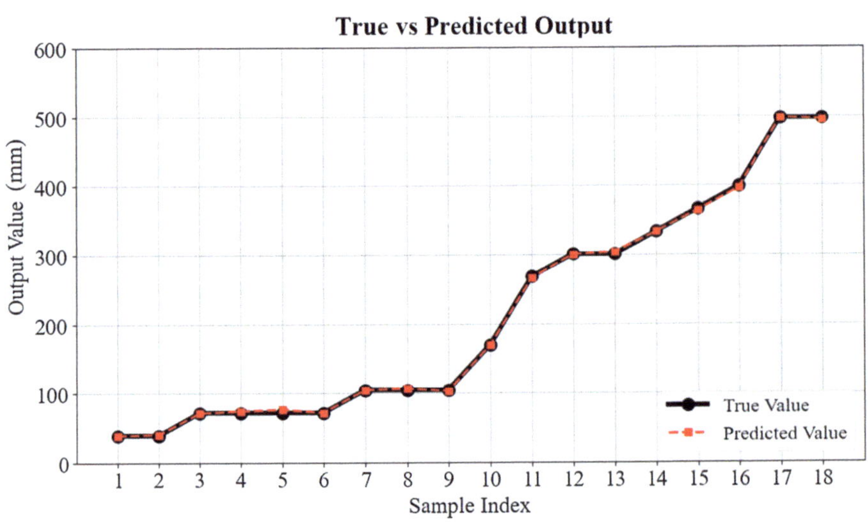

Fig. 27 Cascade-forward neural network internal test set performance

Fig. 28 Cascade-forward neural network residual errors

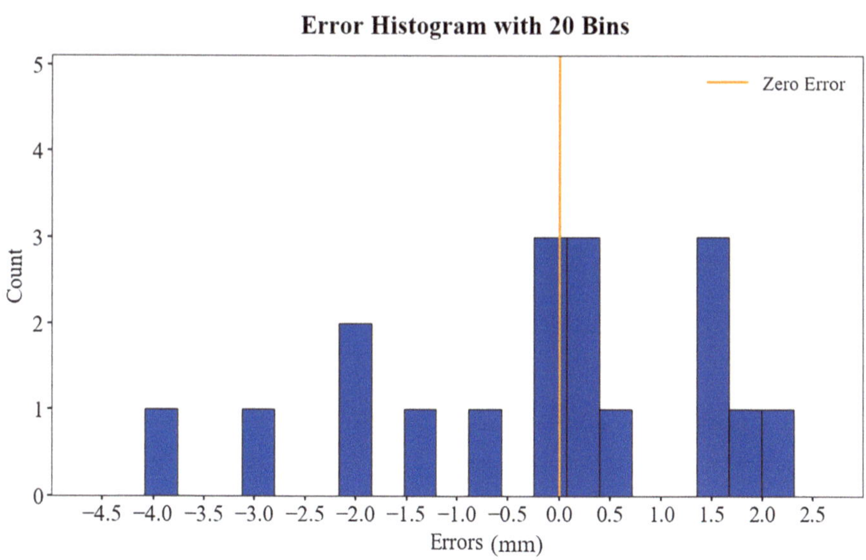

Fig. 29 Cascade-forward neural network error histogram

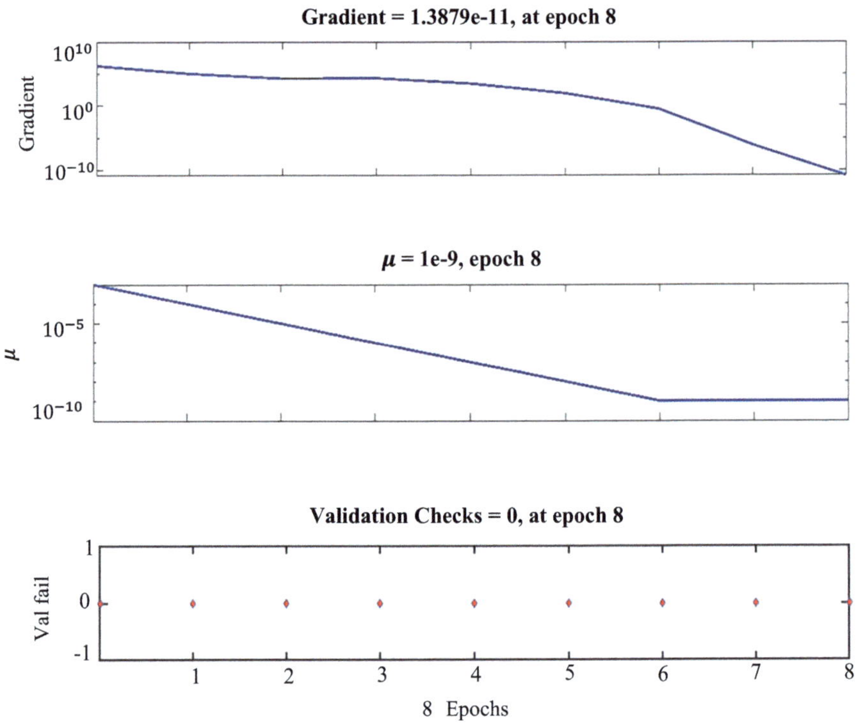

Fig. 30 Cascade-forward neural network training state plot

Table 15 Cascade-forward neural network numerical performance

Performance metrics	Training set	Validation set	Test set
RMSE	0.000 mm	2.4525 mm	1.6978 mm
MAE	0.000 mm	1.8774 mm	1.2818 mm
MAPE	0.00%	1.23%	1.11%
R^2	1.000	0.9996	0.999

Table 16 Cascade-forward external test set performance

True value (mm)	Predicted value (mm)	Absolute error (mm)	Relative error (%)
54.000	55.870	1.8701	3.4632
98.000	96.798	1.2015	1.2260
123.00	133.67	10.671	8.6757
137.00	142.52	5.5249	4.0327
181.00	193.79	12.7860	7.0639
212.00	219.73	7.7274	3.6450
289.00	295.70	6.6990	2.3180
315.00	328.81	13.8130	4.3850
371.00	374.77	3.7741	1.0173

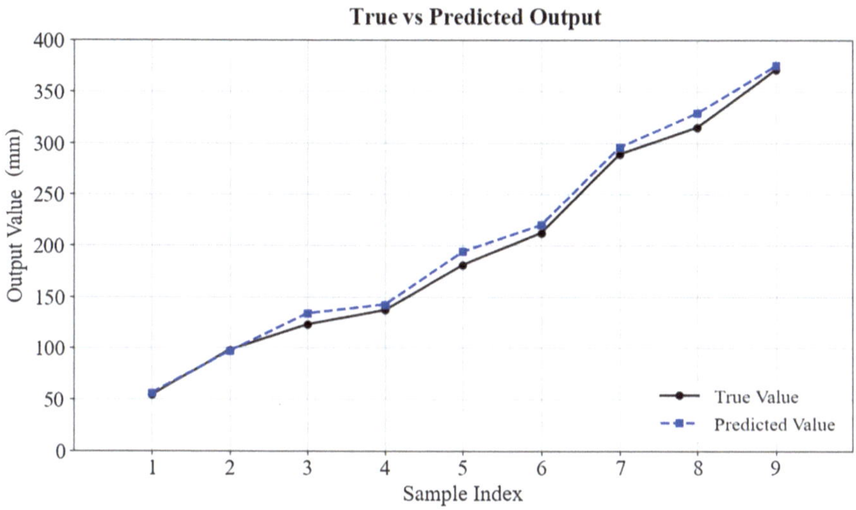

Fig. 31 Cascade-forward neural network external test set performance

Table 17 Cascade-forward axial detection performance

Performance metrics	Training set	Validation set	Test set
RMSE	0.5164	1.4564	1.8330
MAE	0.1333	0.5305	0.8372
MAPE	3.60%	12.00%	51.58%
R^2	0.9477	0.5777	0.4772

Despite the larger errors seen with external testing, these values are still suitable for a broad model used to analyze new defects. The fact that most positions gave relative errors less than 5% shows that our model can estimate well, especially since these samples were not included in the training data.

Axial Detection Performance (Table 17)

The cascade-forward network fits the training data very well (RMSE = 0.52, MAE = 0.13, MAPE = 3.6 %, $R^2 \approx 0.95$), but its performance decreases on unseen data: validation errors rise to RMSE ≈ 1.46, MAE ≈ 0.53 ($R^2 \approx 0.58$), and test errors reach RMSE ≈ 1.83, MAE ≈ 0.84 ($R^2 \approx 0.48$) with a MAPE of ~51.6 %, indicating that, despite strong fitting on the training set, the model still struggles to generalize reliably for precise axial defect localization

Table 18 FitNet Internal test set performance

Axes label	True value (mm)	Predicted value (mm)	Absolute error (mm)	Relative error (%)
1	39.000	38.757	0.24303	0.62316
7	39.000	38.845	0.15537	0.39837
4	71.760	70.840	0.92025	1.2824
5	71.760	72.670	0.91037	1.2686
8	71.760	71.228	0.53171	0.74096
2	71.760	72.204	0.44400	0.61873
1	104.39	104.39	0.000	0.000
4	104.39	106.48	2.0876	1.9998
6	104.39	103.94	0.45425	0.43514
8	169.83	170.73	0.90492	0.53284
3	267.99	266.88	1.1055	0.41252
1	300.71	302.24	1.5309	0.50909
7	300.71	301.68	0.97237	0.32336
2	333.43	333.43	0.000	0.000
4	366.15	363.37	2.7819	0.75977
6	398.87	396.56	2.3104	0.57925
1	497.03	497.54	0.51327	0.10327
3	497.03	497.12	0.09309	0.01873

3.1.4 Fitnet network

This section presents the prediction performance done by the fitnet network in estimating the defect position.

Internal Test Set Performance

Table 18 shows the internal test results, where most errors are below 2% with the highest being 1.99% at 104.39 mm position.

Figure 32 represents the regression plots for training, validation, and test datasets. All three of them are follows the ideal line $y = x$, with R^2 values with 0.9997. This illustrates that there is a good correlation and minimal variance across the splits.

Figures 33 and 34 show the predicted vs actual values along with their residuals. In Fig. 34, where residuals are small and evenly distributed around zero, the model does not show any noticeable bias in its predictions

Figure 35 shows the histogram of the residual errors. Most errors are centered closely around zero, reflecting the consistency of the network's output. Errors seem to tend toward the positive side.

The error histogram shown in Fig. 35 has most of the errors concentrated around zero with a relatively symmetric distribution. The highest we can see tend towards 2.5, as observed most of the errors are on the positive side

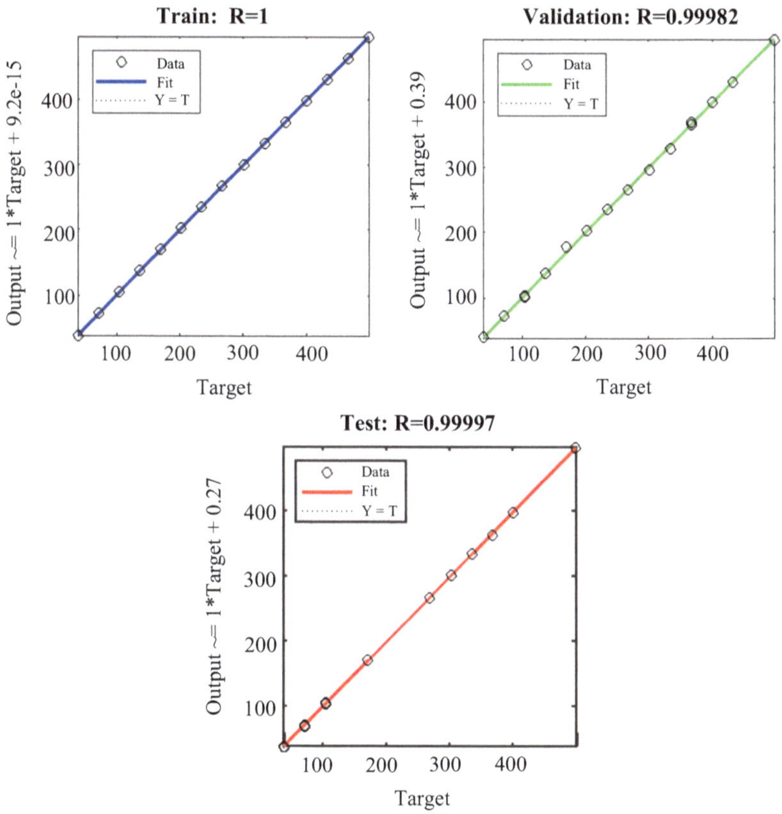

Fig. 32 Fitnet regression plots

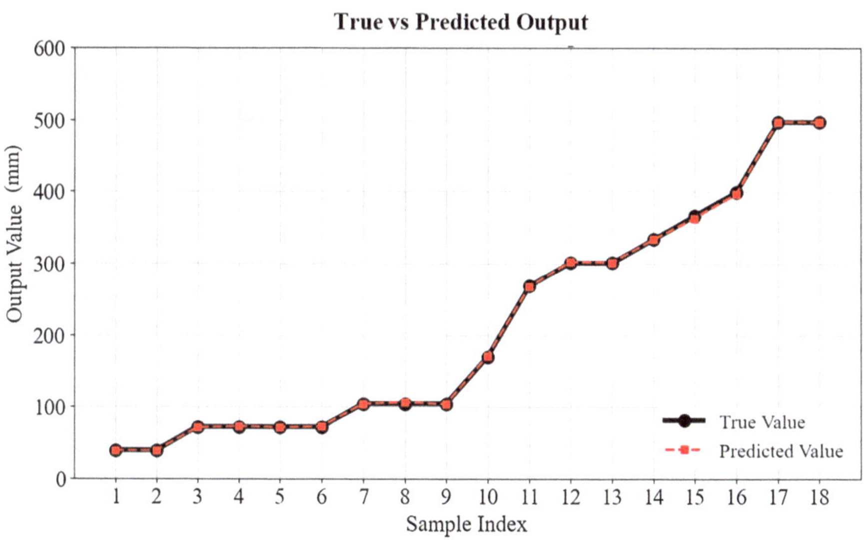

Fig. 33 Fitnet neural network internal test set performance

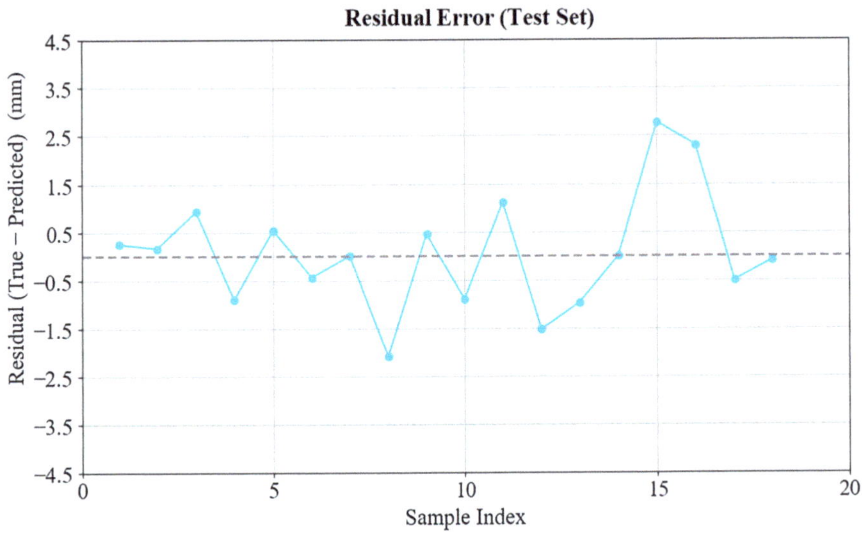

Fig. 34 Fitnet neural network residual errors

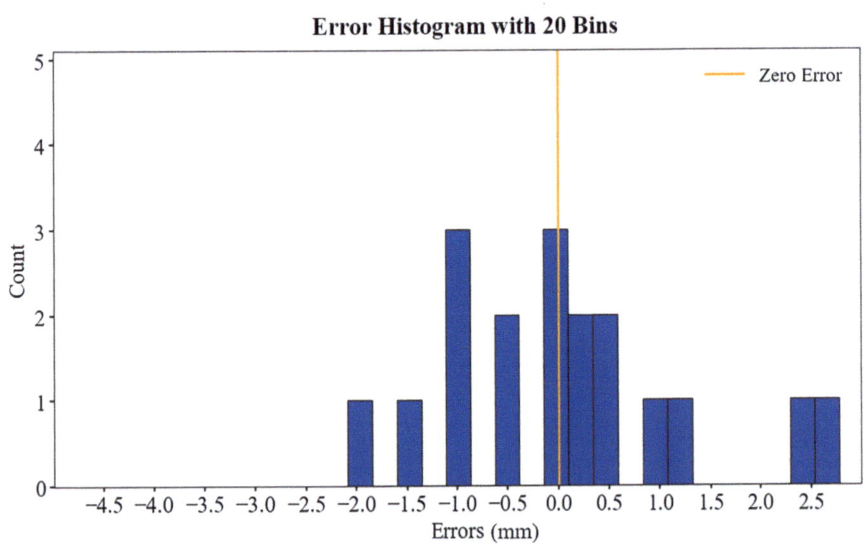

Fig. 35 Fitnet neural network error histogram

Figure 36 shows the training network progression. No validation failures occurred and the model remained stable during training. This suggests that the model understood the training data.

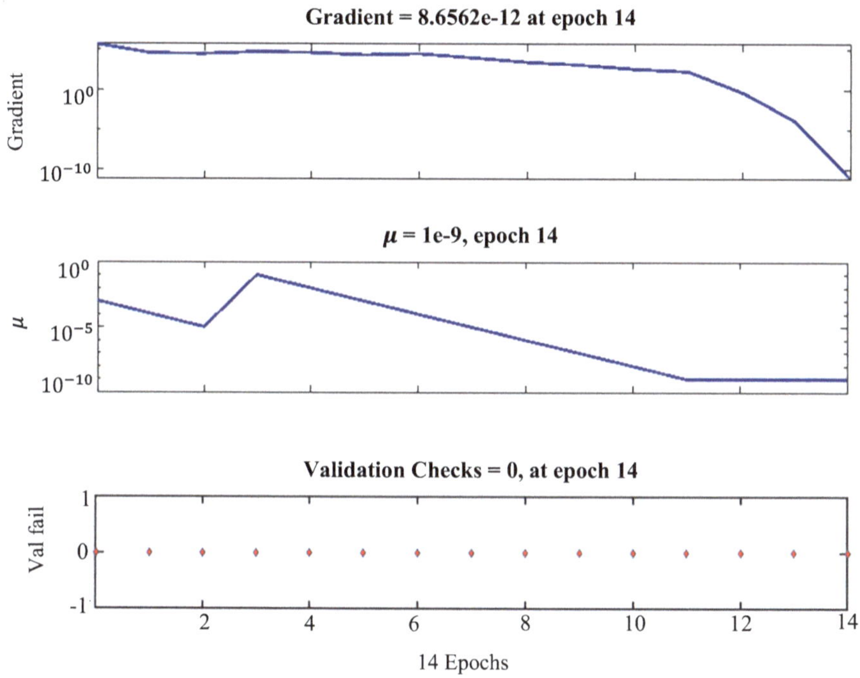

Fig. 36 Fitnet neural network training state plot

Table 19 Fitnet neural network numerical performance

Performance metrics	Training set	Validation set	Test set
RMSE	0.000 mm	2.5035 mm	1.1898 mm
MAE	0.000 mm	1.5279 mm	0.8866 mm
MAPE	0.00%	0.81%	0.59%
R^2	1.000	0.9996	0.999

Figure 36 illustrates the fitnet network's training behavior. By epoch 14, the gradient has fallen to about 8.66×10^{-12}, indicating that the network's error surface has effectively flattened and the training has converged

Table 19 proves that the network demonstrated its capabilities like the rest in learning the fault longitudinal location patterns effectively with minimal error.

External Test Set Performance

Table 20 presents the predicted versus the true defect positions for nine samples of data. As can been seen the largest error recorded was 5.64% at the 123 mm position

Table 20 Fitnet external test set performance

True value (mm)	Predicted value (mm)	Absolute error (mm)	Relative error (%)
54.000	55.538	1.5377	2.8477
98.000	100.84	2.8392	2.8972
123.00	129.94	6.9425	5.6443
137.00	141.48	4.4767	3.2677
181.00	186.52	5.5201	3.0498
212.00	217.55	5.5468	2.6164
289.00	294.21	5.2075	1.8019
315.00	325.72	10.715	3.4016
371.00	389.57	18.565	5.0041

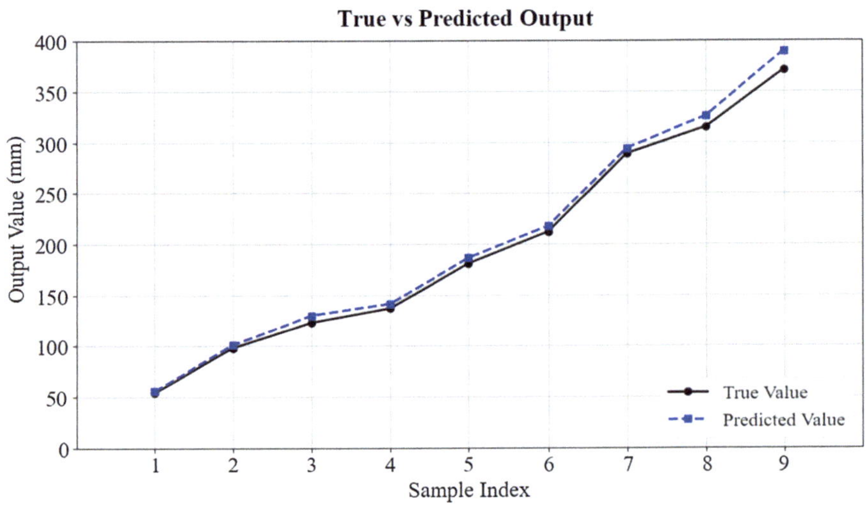

Fig. 37 Fitnet neural network external test set performance

defect, however, in the rest of the cases, the relative error is less than 5%, indicating its good performance.

Figure 37 illustrates the locations of actual and predicted defects. The points usually stay along the identity line but deviate a little more as the pipe gets longer. So, while the model performs well overall, some defects that occur at long ranges may be more easily disrupted by changes in the frequency signature.

The performance metrics for the external test set are as follows:

- MAE: 6.82 mm
- RMSE: 8.34 mm
- MAPE: 3.39%

Table 21 Fitnet axial detection performance

Performance metrics	Training set	Validation set	Test set
RMSE	0.5164	1.3914	1.1221
MAE	0.1333	0.5884	0.5515
MAPE	3.60%	12.62%	18.74%
R^2	0.9477	0.6145	0.8041

Overall, the deviations in some cases were small. Most predictions were only off by a few percent, which demonstrates the model can handle defects it has not seen before.

Axial Detection Performance (Table 21)

The fitnet model also fits the training data very well (RMSE = 0.52, MAE = 0.13, MAPE = 3.6 %, $R^2 \approx 0.95$). On the validation set, errors rise moderately (RMSE ≈ 1.39, MAE ≈ 0.59, MAPE ≈ 12.6 %, $R^2 \approx 0.61$), but on the test set, the performance improves notably (RMSE ≈ 1.12, MAE ≈ 0.55, MAPE ≈ 18.7 %, $R^2 \approx 0.80$), suggesting that fitnet generalizes better than the other architectures and offers more reliable axial defect-localization accuracy.

3.1.5 Linear Regression Model

This section presents the performance results of the linear regression model developed to predict defect locations in steel pipelines based on frequency shift features extracted through modal analysis. Two training-test splits were implemented to investigate how the model's performance responds to different training data proportions:

- Model A: 70% training and 30% testing split
- Model B: 80% training and 20% testing split

The predictive performance of both models is discussed in terms of their numerical accuracy and reliability on their respective test sets. In addition, the capability of the linear regression model to detect angular (circumferential) defect positions is explored using model A, as this split provided sufficient data distribution for evaluating circumferential predictions.

Standard regression metrics — including mean absolute error (MAE), root mean square error (RMSE), mean absolute percentage error (MAPE), and the coefficient of determination (R^2) — are used throughout this section to assess model accuracy and generalization behaviour.

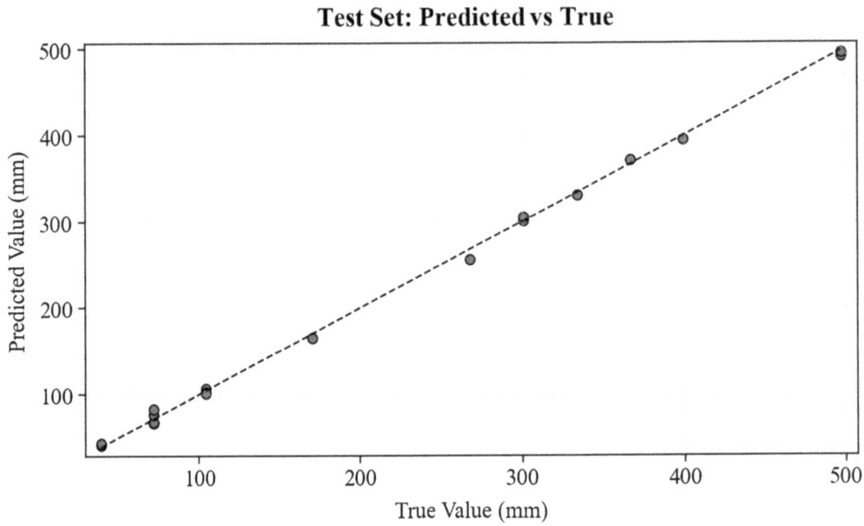

Fig. 38 Linear regression test set performance (70/30)

3.1.6 Model A - 70/30 Split Performance

The predictive capability of the methodology was tested by using a split of 70% data for training and 30% for testing. Because of this strategy, the model learned from most of the data, while still having enough for tests. It was decided to use linear regression for its ability to explain the data because it could detect linear connections between frequency signatures and defects.

The graph at Fig. 38 compares the actual values with the predicted output. A close linear trend in the data suggests the model makes quite accurate predictions. Most of the points are found along the identity line (Y = X), showing that there are very few residuals in the evaluations.

The model training results from Table 22 show that since there was little difference in error between training, the model can handle data similar to what it saw in training. The model's ability to predict well and show low variability in residuals is backed up by the large R^2 values.

Afterward, the model was applied to new data to determine if it could be practical. Test data was 15% of the total and was kept out of training. Figure 39 shows the residual error plot which illustrates the gap between the predicted and real outcomes for each test record.

The results show that the residuals have values mostly close to zero and no explicit pattern which shows that our model is dependable and does not prefer one defect group over another.

Figure 40 shows a histogram of residuals, showing again that they roughly follow a normal distribution.

Table 22 Linear regression numerical performance (70/30 Split)

Dataset	RMSE (mm)	MAE (mm)	MAPE (%)	R^2
Training set	3.9040	3.1202	1.96	0.9992
Test set	5.3405	4.5302	4.13	0.9988

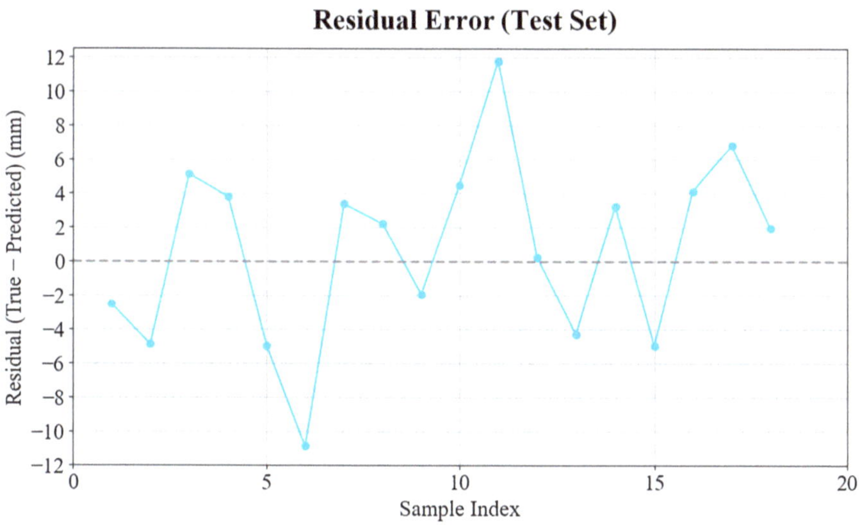

Fig. 39 Linear regression residual error (70/30)

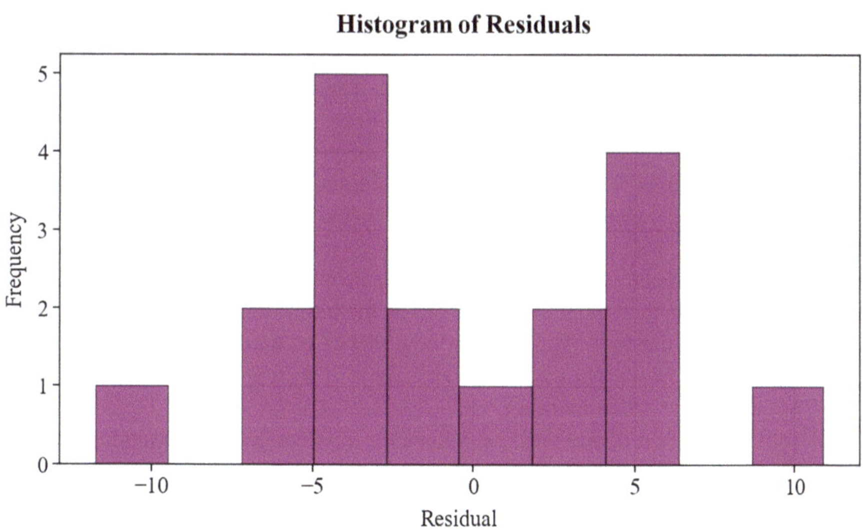

Fig. 40 Linear regression prediction error histogram (70/30)

Table 23 Linear regression test set performance (70/30 Split)

Axes label	True value (mm)	Predicted value (mm)	Absolute error (mm)	Relative error (%)
1	39.0	41.52	2.5201	6.4617
7	39.0	43.88	4.8806	12.514
4	104.39	106.34	1.9482	1.8663
5	104.39	102.2	2.1942	2.1019
8	169.83	165.37	4.4627	2.6277
2	300.71	300.44	0.2671	0.0888
1	333.43	330.22	3.2085	0.9623
4	366.15	371.13	4.9757	1.3589
6	398.87	394.76	4.1108	1.0306
8	497.03	495.06	1.9661	0.3956

Since the histogram is symmetric, this means randomness in the errors which is essential for a good prediction model.

For a clearer picture of model accuracy, Table 23 provides the first 10 test predictions, actual numbers and the corresponding errors. The difference between the actual alignment and the predicted alignment is close, with most relative errors below 3%, except when the ground truth is very small.

Model A – Generalization Performance

To evaluate generalization, model A was tested on nine unseen pipe configurations with defect locations not included in the training set. This test is essential in predictive maintenance, where models must identify defects in new, untrained conditions. Each pipe contained a single circular hole placed at unique axial coordinates, excluded from the original defect grid. The material, hole dimensions, and boundary conditions were kept consistent with the training data. Natural frequencies were extracted using the same modal analysis procedure described in Sect. 2. Frequency signatures were calculated by subtracting the control pipe's modes from the test pipes, and the model then predicted the defect location based on these new inputs.

Table 24 provides that the MAE and RMSE from the test set that the predictions were slightly worse than what the model did on the training set. This outcome is to be expected because the tests were run on new pieces of hardware. The fact that the MAPE stayed below 6% showed the model was still accurate in predicting locations it did not encounter when learning.

In Table 25, the predicted and true values for the nine generalization test sample, as well as their absolute and relative errors are given:

Most of the predicted values fall within acceptable errors, but errors are higher when predicting 137 and 54. They are likely to occur because the model is called on

Table 24 Linear regression generalization test performance (70/30 Split)

MAE (mm)	MAPE (%)	RMSE (mm)	No. of Samples
8.5843	5.5054	10.7749	9

Table 25 Linear regression prediction error summary (70/30 Split)

True value (mm)	Predicted value (mm)	Absolute error (mm)	Relative error %
54	46.713	7.2867	13.494
98	99.555	1.5555	1.5872
123	122.95	0.04808	0.03909
137	155.38	18.381	13.417
181	195.78	14.785	8.1685
212	221.59	9.5874	4.5223
289	297.16	8.1589	2.8231
315	331.54	16.541	5.251
371	370.08	0.91579	0.24684

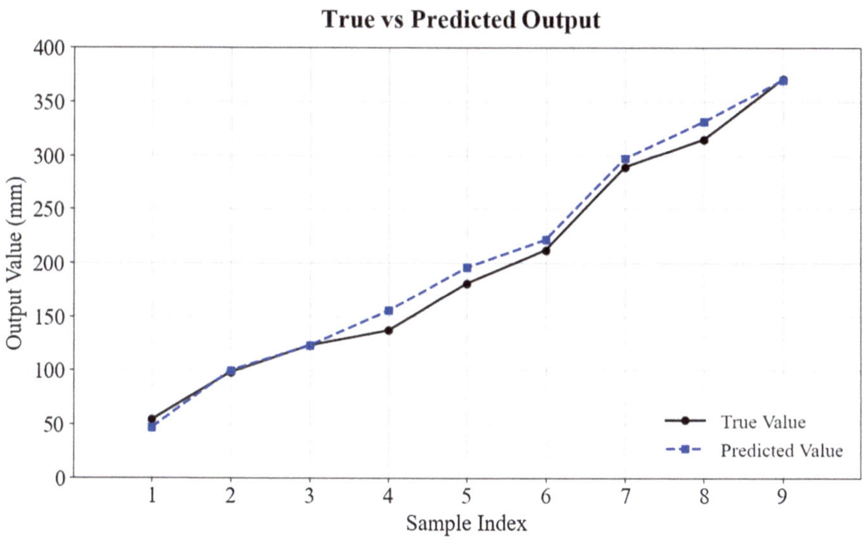

Fig. 41 Linear regression generalization test performance (70/30)

to predict data points that are far from the dataset's most dense areas. On the other hand, the low error seen in 123 and 371 seems to indicate the model is still effective when the frequency pattern is like those from the training set.

Figure 41 shows the comparison of the actual pipe results with the model's performance when given completely new data. In the figure, each point on the chart indicates how close or far the model's prediction is from the correct answer.

Most of the data points are found close to the 45-degree identity line which means the model still performs well without learning from these positions. Yet, the model deviates more at the lower and higher extremes of the range, for example 54 and 137. Given the nature of the data, examples that come from the extreme edges are challenging for the model because they lie far from the bulk of the training examples. However, the model continues to maintain accuracy for predicting defects mainly along the identity line and when it deals with new configurations.

Model A - Angular Detection

The section shows the outcomes of using a linear regression model to predict the angular position of faults along the pipe based on vibration frequency features. The purpose of this analysis was to test whether the model could accurately identify the circumferential location of defects. A 70/30 data split was used in this experiment, as the limited number of angular positions in the dataset made the 80/20 split impractical for reliable evaluation. The 70/30 split ensured a more balanced distribution of training and testing samples, allowing for meaningful performance

analysis. However, the model failed to capture the relationship between the input features and angular positions, as indicated by a negative R^2 value, showing that the linear regression approach was not effective for this task (Table 26).

As given in Table 27, the results indicate the model had trouble accurately forecasting where the defects are located. Some predictions were accurate, but the overall trend showed big mistakes and poor connection with the real results, mainly on the test data.

Table 26 Linear regression axial detection performance

Dataset	RMSE	MAE	R^2
Training set	2.1921	1.9109	0.0847
Test set	2.3855	2.1235	−0.0839

Table 27 Linear regression axial prediction error summary

True value	Predicted value	Absolute error	Relative error (%)
1	2.2487	1.2487	124.87
6	4.0583	1.9417	32.36
7	2.1263	4.8737	69.62
2	4.6057	2.6057	130.29
5	4.4990	0.5010	10.02
7	4.4769	2.5231	36.04
3	5.0015	2.0015	66.72
3	4.6397	1.6397	54.66
7	4.6050	2.3950	34.21
2	4.8307	2.8307	141.53

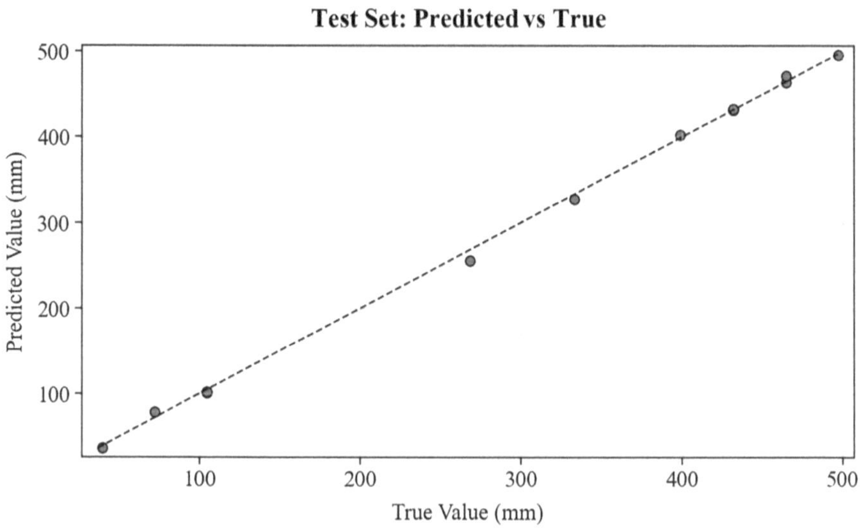

Fig. 42 Linear regression test set performance (80/20)

Table 28 Linear regression numerical performance (80/20 Split)

Dataset	RMSE (mm)	MAE (mm)	MAPE (%)	R^2
Training Set	4.1172	3.2825	2.51	0.9992
Test Set	5.0585	3.8475	2.58	0.9991

3.1.7 Model B - 80/20 Split Performance

A second linear regression model called model B was created using an 80/20 train// test split to look at the impact of training volume. Here, the training data was made up of 80% of the data and 20% for testing. The aim was to study whether expanding training data helps the model generalize better and if using small test sets could alter its performance.

The training phases of model B showed strong predictive performance, much like the earlier model with a 70/30 split. In fact, there were some improvements in marking.

Figure 42 presents the predicted vs. true scatter plot for model b across the training sets.

The performance of model B was slightly higher than that of model A, as given in Table 28, mainly because of the few errors it made in MAE and MAPE. Even though this is a modest enhancement, it is still worth pointing out, since the accuracy holds steady even with the smaller number of validators.

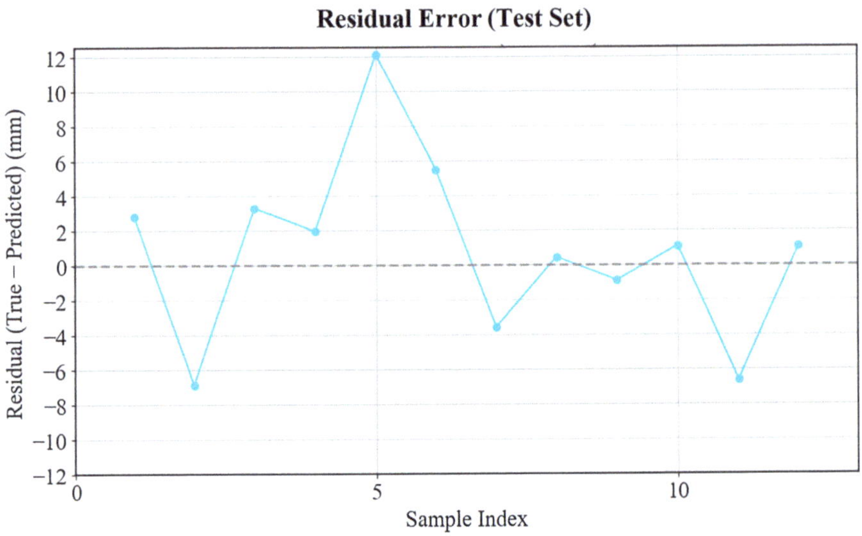

Fig. 43 Linear regression residual error (80/20)

Table 28 provides that model B accurately evaluated cases that were not included in its training set. The way errors were handled stayed very close to the desired ranges

The residual error plot is shown in Fig. 43 for the test set predictions. The symmetry of residuals shows that the error structure is well balanced.

In Fig. 44, the histogram of residuals, appears bell-shaped, suggesting that linear regression can't be rejected for having normally distributed residuals.

This model is illustrated in Table 29, where we see how it made predictions on the first ten test samples. Most of the time, the relative errors are less than 3%. This validates what the numbers have told us and what we have seen in the images.

Just as with the earlier model, smaller values produce slightly higher errors, as even a little bit of deviation makes a bigger percentage difference. The model's predictions are still judged to be useful from a diagnostic engineering viewpoint.

Model B – Generalization Performance

To evaluate model B, the performance of the same nine pipeline setups was chosen to evaluated how well it generalizes. The mean location has been used separately from all training and test datasets to mimic real-world flaws the model may not know about. The generalizations used in both models made it possible to directly compare model B and model A in new and different situations.

Table 30 explains the generalization performance of the 80/20 model. The metrics from the external test set reveal a minor rise in error, as expected, but are still within the limits typically accepted for diagnosis.

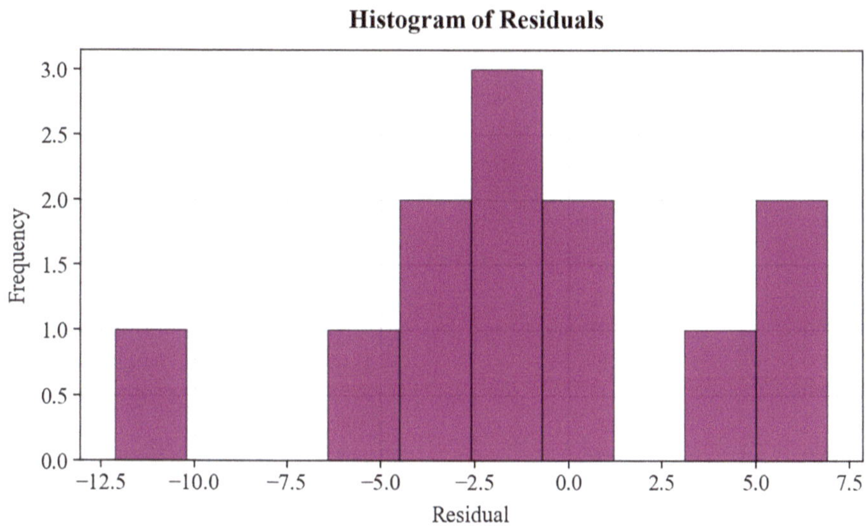

Fig. 44 Linear regression prediction error histogram (80/20)

Table 29 Linear regression test set performance (80/20 Split)

Axes Label	True Value (mm)	Predicted Value (mm)	Absolute Error (mm)	Relative Error (%)
1	39.0	36.213	2.7866	7.1452
7	71.76	78.668	6.9085	9.6272
4	104.39	101.12	3.2692	3.1318
5	267.99	255.88	12.11	4.519
8	333.43	327.94	5.4868	1.6456
2	398.87	402.46	3.5867	0.89922
1	431.59	432.5	0.90557	0.20982
4	431.59	431.16	0.43075	0.09981
6	464.31	463.25	1.0646	0.22929
8	464.31	470.97	6.6563	1.4336

Table 30 briefly explains how the 80/20 model carried out generalization. The external test set led to metrics that are a little higher in error but remain acceptable for diagnostic purposes.

Absolute and relative errors for the nine generalization samples are given in Table 30, where they are compared to the correct values. Outcomes generally within +/−10% of the actual results are remarkable since these situations are difficult for the model to cope with (Table 31).

The biggest error took place at the ends (54 and 137), because there were few examples during model training at those points. On the other hand, the developed program showed a nearly perfect prediction for 123 and 371. The outcomes from

Table 30 Linear regression generalization test performance (80/20 Split)

MAE	MAPE (%)	RMSE	No. of samples
8.6589	5.7766	10.7794	9

Table 31 Linear regression prediction error summary (80/20 Split)

True value (mm)	Predicted value (mm)	Absolute error (mm)	Relative error (%)
54.0	46.38	7.61980	14.1110
98.0	101.45	3.45090	3.52140
123.0	123.33	0.33295	0.27069
137.0	156.32	19.3210	14.1030
181.0	195.51	14.5110	8.01700
212.0	221.11	9.10540	4.29500
289.0	296.1	7.10080	2.45700
315.0	331.07	16.0720	5.10230
371.0	370.58	0.41592	0.11211

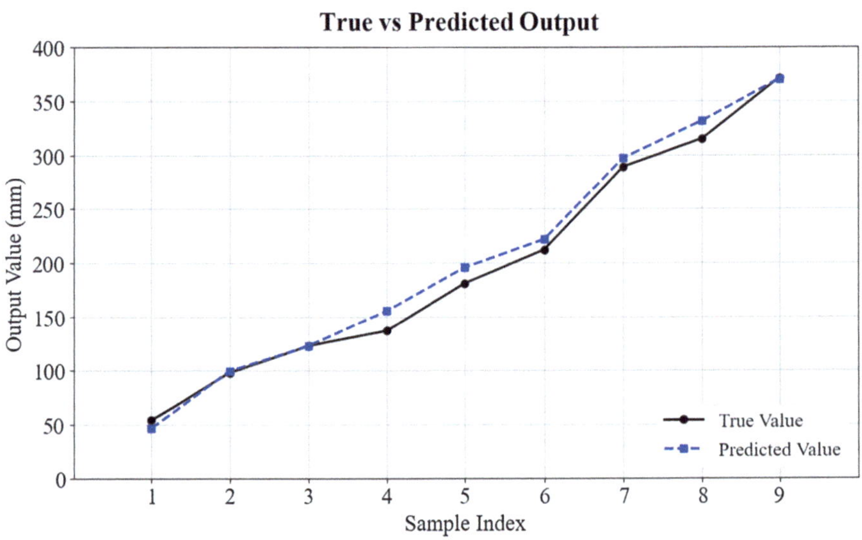

Fig. 45 Linear regression generalization test performance (80/20)

model B confirm that the patterns produced by the new data also lie close to the decision boundary set by model A.

The real versus predicted comparison is shown for each generalization sample in Fig. 45 using model B. The points align well with the identity line, meaning the model continues to offer accurate results for all kinds of defects.

The trend continues to be linear and centered on the true values, even though the results are less tightly concentrated than those seen in internal tests, with no bias toward over- or under-prediction.

3.1.8 ANN Performance Summary

This section will compare all three networks: feedforward, cascade-forward and fitnet models in terms of their ability to predict defect positions with data from low-frequency modal analysis. The focus of the comparison is on how accurate predictions are made, how the model generalizes.

Internal Test Set Performance Summary

This section summarizes the performance of the three networks in predicting the longitudinal defect positions for the internal test (Table 32).

Each of the three models did very well in the internal test, with R^2 values greater than 0.999 and MAPE results below 1.2%. The feedforward network got the best results, with a MAE of 0.54 mm and an RMSE of 0.79 mm. While there were slightly more errors seen in the cascade model, it still maintained high consistency with minimal variance.

Overall, the three networks were successful in learning the frequency shift associates with the defect longitudinal position and predicting it.

External Test Set Performance Summary

This section summarizes the performance of the three networks in predicting the longitudinal defect positions for the external test (Table 33).

All three models showed a drop in accuracy when tested on unseen data, which was as we expected. The fitnet model showed a slightly lower MAPE (3.39%) between the three.

Figure 46 compares the absolute error from their external performance test between all three networks. The Cascade-forward network shows moderately low errors at the early detection position, but as the defect position increases, there is a

Table 32 Comparison of neural networks model performance on internal test set

Model	MAE (mm)	RMSE (mm)	MAPE (%)	R^2
Feedforward	0.54	0.79	0.47	0.9993
Cascade-Forward	1.28	1.70	1.11	0.9999
Fitnet	1.03	1.48	0.94	0.9996

Table 33 Comparison of neural networks model performance on external test set

Model	MAE (mm)	RMSE (mm)	MAPE (%)
Feedforward	9.09	7.53	4.55
Cascade-Forward	7.12	8.32	3.98
Fitnet	6.82	8.34	3.39

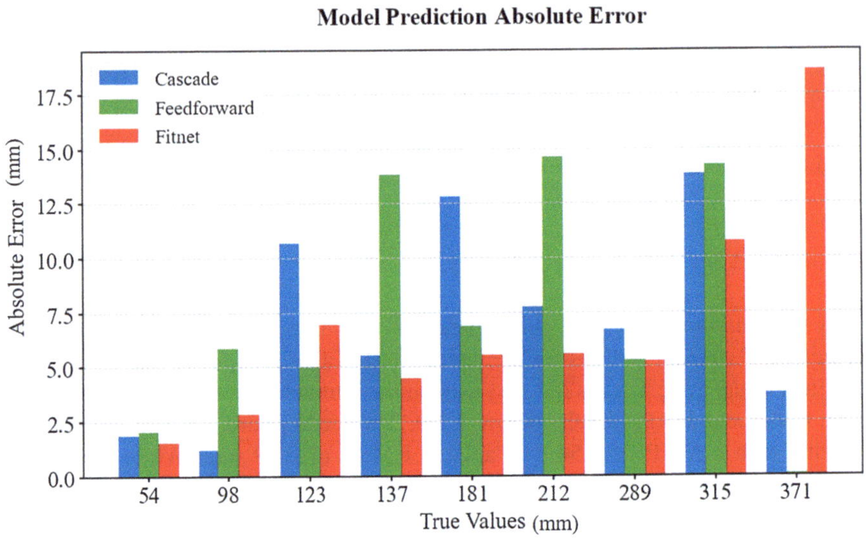

Fig. 46 Comparison of neural networks model prediction absolute error

Table 34 ANN axial detection performance summary

Model	RMSE	MAE	MAPE (%)	R^2
Feedforward	1.75	0.86	47.00	0.52
Cascade-Forward	1.83	0.84	51.58	0.48
Fitnet	1.12	0.55	18.74	0.80

gradual increase in the errors. The feedforward network performs well at low defect positions, but some error spikes at mid to late positions, indicating inconsistent performance overall. The fitnet network has the lowest error except at position 371 mm, where it shows a dramatic spike. Otherwise, the model is stable and accurate in this comparison.

The results suggest that the fitnet model performed the best, where it dealt well with unseen defects, which is vital for practical pipe inspection. Feedforward performance is the best inside the model's learned domain, yet fitnet's reliable performance outside this domain makes it the easiest to deploy.

Axial Detection Performance Summary (Table 34)

Fitnet shows the best results overall, since its predictions are very close to the actual circumferential axes locations of the defects. Feedforward performs better than Cascade-Forward which has the poorest generalization ability. The results prove that fitnet is the best architecture for tracking and pinpointing defects.

In summary, feedforward and cascade-forward neural networks were showing signs of overfitting, where the model starts memorizing. However, fitnet showed some good promises having the smallest generalization gap and the best test-set accuracy (training vs. validation). Between the three, fitnet emerges as the most promising, since its test $R^2 \approx 0.80$ and MAPE $\approx 18.7\ \%$ indicate a level of precision much closer to acceptable than the roughly 50 % error seen in the other two.

A major reason for the inaccuracy in the test set across the three networks is that the effects of defects in the circumferential axes on the natural frequency shifts are small. This is mainly due to the defects large size in comparison to the pipe dimensions. Where certain defects exhibit identical frequency shifts, the network ends up latching onto tiny noise patterns in the training set that don't generalize. In addition, the perfect symmetric pipe means defect mirrored locations frequency shifts can look almost the same, making it difficult to learn robust distinguishing features without memorizing the training examples.

3.1.9 Linear Regression Performance Summary

A direct comparison between model A (70/30 split) and model B (80/20 split) highlights their respective strengths and the impact of increased training data.

Performance Summary (Table 35)

The fact that MAPE and MAE declined for model B suggests the added data in training improved how consistently the model could predict the values.

Generalization Performance Summary (Table 36)

These findings indicate that, though model B worked well, adding more data during training did not substantially improve handling of completely new situations.

Both models showed accurate results, unbiased data and good generalization. The training data for the test set was more abundant for model B compared to model A but model A performed slightly better in non-trained situations. The results prove

Table 35 Linear regression training performance comparison

Model	RMSE	MAE	MAPE (%)	R^2
Model A	5.3405	4.5302	4.13	0.9988
Model B	5.0585	3.8475	2.58	0.9991

Table 36 Linear regression generalization test comparison

Model	RMSE	MAE	MAPE (%)
Model A	10.7749	8.5843	5.5054
Model B	10.7794	8.6589	5.7766

that using frequency signature features with linear regression is valuable for making predictions about defects in pipelines.

Axial Detection Performance Summary

The performance of the linear regression model in predicting the axial (circumferential) position of defects was found to be unsatisfactory. Using the 70/30 data split, the model was trained and tested on frequency signature inputs associated with eight angular positions. The results indicate that the model had difficulty accurately forecasting where the defects were located based on frequency domain information. While a few predictions were close to the true positions, the overall trend showed large deviations and a poor relationship with the actual defect locations, particularly in the test data.

This weak performance is likely due to the limited variation in frequency signatures between neighboring angular positions. The angular defects were relatively large and closely spaced, leading to similar vibration responses that the linear model could not distinguish. As a result, the linear regression model failed to capture meaningful patterns in the data. This outcome suggests that fine angular localization is difficult when physical differences between defect cases are minimal. Future improvements could involve reducing defect sizes, increasing the angular spacing between defects, or utilizing nonlinear models better suited to capturing small-scale variations in vibration response.

4 Conclusion and Future Work

This research presents a vibration-based approach for detecting defects in pipeline structures using modal analysis combined with machine learning techniques, namely artificial neural networks (ANNs) and linear regression. A total of 130 pipe models were simulated in ANSYS, comprising 129 pipes with single-hole defects at various locations and one defect-free control pipe. The first 15 natural frequencies from modal analysis were extracted and used as input features to develop models capable of predicting defect positions.

Three ANN architectures—feedforward, cascade-forward, and fitnet—were trained and evaluated. Among these, the fitnet model showed the highest accuracy in predicting longitudinal defect locations, with a mean absolute percentage error (MAPE) of less than 1% and strong regression performance, particularly on unseen test data. However, attempts to predict circumferential positions were less successful. Large defect sizes produced nearly identical frequency shift signatures across mirrored axes, limiting the models' ability to differentiate angular locations. Smaller and more spaced-out defects might be more distinguishable, and this remains an opportunity for future improvement.

To complement the ANN models, a linear regression model was developed using the same dataset and structure, with two separate data splits (Model A: 70/30 and Model B: 80/20) to assess how training size affects performance. While the linear regression models offered interpretability and a useful baseline, their prediction accuracy was notably lower than that of the ANN models, especially for circumferential detection. In particular, the 70/30 split was applied to test angular defect localization, but the model returned a negative R^2 value, indicating poor predictive capacity due to the similarity of vibration patterns across neighboring angles. This limitation highlights the complexity of capturing subtle variations in structural response with simple linear models.

Despite these challenges, both modeling approaches reinforced the feasibility of using low-frequency modal data for fault detection in pipelines. This work contributes a simulation-based pipeline fault detection framework that is non-invasive, data-efficient, and suitable for future real-time monitoring applications. To validate real-world implementation, the proposed system can be tested experimentally by exciting pipelines with a shaker or mechanical vibrator for 10 s and capturing response data with accelerometers. These signals can then be fed to the trained models for defect localization.

Future work should focus on increasing dataset size, testing additional defect geometries, and incorporating experimental data. Modeling smaller defects (e.g., <2 mm in diameter) or increasing the angular spacing between holes could enhance angular detection. Furthermore, advanced nonlinear models or hybrid AI techniques may offer better performance in capturing the subtle vibration signatures associated with small or closely spaced defects. With further development, this framework has the potential to become a robust tool for non-destructive evaluation (NDE) and structural health monitoring (SHM) in industrial pipelines.

Appendices

Appendix A: MATLAB Code for Data Preparation

```
%% Parameters
inputFilename = 'pipe_fault_signatures(in).csv'; % Name of the master
CSV file
labelColumn = 'zone_label'; % Name of the column containing axial-
position indices (1-15)

% Proportions for splitting
pTrain = 0.70;
pVal = 0.15;
pTest = 0.15; % (implicitly 1 - (pTrain+pVal))

rngSeed = 42; % Fixed seed for reproducibility
```

```
%% Read data
% Use readtable so that column names are preserved
T = readtable(inputFilename);

% Extract the axial-position labels (1 through 15)
if ismember(labelColumn, T.Properties.VariableNames)
 axialPosLabels = T.(labelColumn);
else
 error('Column "%s" not found in "%s". Check the column name or index.',
labelColumn, inputFilename);
end

N = height(T);
if numel(axialPosLabels) ~= N
 error('Length of axialPosLabels (%d) does not match number of rows in
table (%d).',...
 numel(axialPosLabels), N);
end

%% Preallocate index arrays
trainIdx = false(N,1);
valIdx = false(N,1);
testIdx = false(N,1);

%% Stratified splitting by axial position
rng(rngSeed);
uniqueGroups = unique(axialPosLabels);

for i = 1:numel(uniqueGroups)
 grp = uniqueGroups(i);

 % Find all row indices for this axial-position group
 grpIndices = find(axialPosLabels == grp);
 nGrp = numel(grpIndices);

 % Determine counts for train/val/test (round to nearest integer)
 nTrain_grp = round(nGrp * pTrain);
 nVal_grp = round(nGrp * pVal);
 % Ensure total sums to nGrp
 nTest_grp = nGrp - nTrain_grp - nVal_grp;

 % Randomly permute these group indices
 perm = grpIndices(randperm(nGrp));
```

```
 % Assign to train/val/test
 trainIdx(perm(1: nTrain_grp)) = true;
 valIdx(perm(nTrain_grp + (1: nVal_grp))) = true;
 testIdx(perm(nTrain_grp + nVal_grp + (1: nTest_grp))) = true;
end

% (Optional) Shuffle boolean masks so that exported rows are not grouped
by axial index
trainIdx = trainIdx(randperm(N));
valIdx = valIdx(randperm(N));
testIdx = testIdx(randperm(N));

%% Create subset tables
T_train = T(trainIdx,:);
T_val = T(valIdx,:);
T_test = T(testIdx,:);

%% Export to CSV
writetable(T_train, 'axial_train_set.csv');
writetable(T_val, 'axial_val_set.csv');
writetable(T_test, 'axial_test_set.csv');

%% Display summary
fprintf('Total samples: %d\n', N);
fprintf(' Training: %d samples --> saved to "train_data.csv"\n', height
(T_train));
fprintf(' Validation: %d samples --> saved to "val_data.csv"\n', height
(T_val));
fprintf(' Test: %d samples --> saved to "test_data.csv"\n\n', height
(T_test));

% Verify each group's representation in each subset
fprintf('Group representation (axial index: train / val / test):\n');
for i = 1:numel(uniqueGroups)
 grp = uniqueGroups(i);
 cntTrain = sum(T_train.(labelColumn) == grp);
 cntVal = sum(T_val.(labelColumn) == grp);
 cntTest = sum(T_test.(labelColumn) == grp);
 fprintf(' %2d: %2d / %2d / %2d\n', grp, cntTrain, cntVal, cntTest);
end
```

Appendix B: MATLAB Code for Training

```
% -----------------------------------------------------
-------------------
% Appendix B: Neural Network Training (Single Run)
% ------------------------------------------
-----------------------------

clear; clc;

%% Load data
trainData = readmatrix('train_set.csv'); % Assumes columns 1-15 are
features, last column is label
valData = readmatrix('val_set.csv');
testData = readmatrix('test_set.csv');

X_train = trainData(:, 1:15); Y_train = trainData(:, end);
X_val = valData(:, 1:15); Y_val = valData(:, end);
X_test = testData(:, 1:15); Y_test = testData(:, end);

%% Normalize features (Z-score)
mu = mean(X_train, 1);
sigma = std(X_train, 0, 1);
sigma(sigma == 0) = 1; % Avoid division by zero if any feature is constant

X_train_norm = (X_train - mu) ./ sigma;
X_val_norm = (X_val - mu) ./ sigma;
X_test_norm = (X_test - mu) ./ sigma;

%% Create Feedforward Network
hiddenLayerSizes = [10 10]; % Two hidden layers, each with 10 neurons
net = feedforwardnet(hiddenLayerSizes, 'trainlm');

% -------------------------------
% Cascade-Forward Network
% -------------------------------
% net = cascadeforwardnet([10 10], 'trainlm');
% % (rest of code is identical: normalization, training, evaluation,
etc.)

% -------------------------------
% FitNet (Function Fitting Network)
% -------------------------------
```

```
% net = fitnet([10 10], 'trainlm');
% % (rest of code is identical)

net.divideFcn = 'dividetrain'; % Disable built-in data division
net.trainParam.showWindow = false; % Turn off training GUI

% Set training parameters (leave min_grad at default 1e-6 unless
otherwise needed)
% net.trainParam.min_grad = 1e-6; % Uncomment to override default
% net.trainParam.epochs = 1000; % Max number of epochs (default is 1000)
% net.trainParam.goal = 0; % Performance goal (leave zero to disable)

%% Train Network (Single Run)
[net, tr] = train(net, X_train_norm', Y_train');

%% Predict on Each Split
Y_train_pred = net(X_train_norm')';
Y_val_pred = net(X_val_norm')';
Y_test_pred = net(X_test_norm')';

%% Compute Performance Metrics
eval_metrics = @(Y, Y_hat) struct(...
 'RMSE', sqrt(mean((Y - Y_hat).^2)),...
 'MAE', mean(abs(Y - Y_hat)),...
 'MAPE', mean(abs((Y(Y~=0) - Y_hat(Y~=0))./Y(Y~=0))) * 100,...
 'R2', 1 - sum((Y - Y_hat).^2) / sum((Y - mean(Y)).^2)...
);

metrics_train = eval_metrics(Y_train, Y_train_pred);
metrics_val = eval_metrics(Y_val, Y_val_pred);
metrics_test = eval_metrics(Y_test, Y_test_pred);

%% Display Metrics
fprintf('\n Training Set Performance:\n');
fprintf(' RMSE: %.4f mm\n', metrics_train.RMSE);
fprintf(' MAE: %.4f mm\n', metrics_train.MAE);
fprintf(' MAPE: %.2f %%\n', metrics_train.MAPE);
fprintf(' R²: %.4f\n', metrics_train.R2);

fprintf('\n Validation Set Performance:\n');
fprintf(' RMSE: %.4f mm\n', metrics_val.RMSE);
fprintf(' MAE: %.4f mm\n', metrics_val.MAE);
fprintf(' MAPE: %.2f %%\n', metrics_val.MAPE);
fprintf(' R²: %.4f\n', metrics_val.R2);
```

```
fprintf('\n Test Set Performance:\n');
fprintf(' RMSE: %.4f mm\n', metrics_test.RMSE);
fprintf(' MAE: %.4f mm\n', metrics_test.MAE);
fprintf(' MAPE: %.2f %%\n', metrics_test.MAPE);
fprintf(' R²: %.4f\n', metrics_test.R2);

%% Extract and Save Weights and Biases
W1 = net.IW{1,1}; b1 = net.b{1};
W2 = net.LW{2,1}; b2 = net.b{2};
W3 = net.LW{3,2}; b3 = net.b{3};
save('Axial_Feedforward.mat', 'net', 'mu', 'sigma', 'tr', 'W1', 'b1',
'W2', 'b2', 'W3', 'b3');

fprintf('\n Model and parameters saved to Axial_Feedforward.mat\n');
```

Appendix C: MATLAB Code for Linear Regression

Data Split and Normalization (70/30 Split)

```
%% Step 1: Load full dataset
data = readmatrix('pipe_fault_signatures.csv');
X = data(:, 1:15);
Y = data(:, end);

n = size(X, 1);

% Initialize index masks
trainIdx = false(n, 1);
testIdx = false(n, 1);

% Stratify based on unique target values
unique_labels = unique(Y);
rng(42); % For reproducibility

for i = 1:numel(unique_labels)
 label = unique_labels(i);
 label_indices = find(Y == label);
 num_label = numel(label_indices);

 % 70% for training
 n_train = round(0.7 * num_label);
 permuted = label_indices(randperm(num_label));
```

```
 trainIdx(permuted(1:n_train)) = true;
 testIdx(permuted(n_train+1:end)) = true;
end

%% Step 2: Extract raw splits
X_train_raw = X(trainIdx,:);
Y_train = Y(trainIdx);

X_test_raw = X(testIdx,:);
Y_test = Y(testIdx);

%% Step 3: Normalize using training set statistics
mu = mean(X_train_raw);
sigma = std(X_train_raw);

X_train = (X_train_raw - mu)./ sigma;
X_test = (X_test_raw - mu)./ sigma;

%% Step 4: Save splits and normalization parameters
save('Stratified_Splits_70_30.mat', 'X_train', 'Y_train', 'X_test',
'Y_test', 'mu', 'sigma');
```

Model Training and Testing (70/30 Split)

```
clear; clc;

%% Step 1: Load stratified 70/30 data and normalization parameters
load('Stratified_Splits_70_30.mat', 'X_train', 'Y_train', 'X_test',
'Y_test', 'mu', 'sigma');

%% Step 2: Train linear regression model
model70 = fitlm(X_train, Y_train);

%% Step 3: Predict on test set
Y_pred = predict(model70, X_test);

%% Step 4: Save model and parameters
save('model_stratified_70.mat', 'model70', 'X_train', 'Y_train',
'X_test', 'Y_test', 'mu', 'sigma');
```

Data Split and Normalization (80/20 Split)

```
%% Step 1: Load full dataset
data = readmatrix('pipe_fault_signatures.csv');
X = data(:, 1:15);
Y = data(:, end);

n = size(X, 1);

% Initialize index masks
trainIdx = false(n, 1);
testIdx = false(n, 1);

% Stratify based on unique target values
unique_labels = unique(Y);
rng(42); % For reproducibility

for i = 1:numel(unique_labels)
 label = unique_labels(i);
 label_indices = find(Y == label);
 num_label = numel(label_indices);

 % 80% for training
 n_train = round(0.8 * num_label);
 permuted = label_indices(randperm(num_label));

 trainIdx(permuted(1:n_train)) = true;
 testIdx(permuted(n_train+1:end)) = true;
end

%% Step 2: Extract raw splits
X_train_raw = X(trainIdx,:);
Y_train = Y(trainIdx);

X_test_raw = X(testIdx,:);
Y_test = Y(testIdx);

%% Step 3: Normalize using training set statistics
mu = mean(X_train_raw);
sigma = std(X_train_raw);

X_train = (X_train_raw - mu)./ sigma;
X_test = (X_test_raw - mu)./ sigma; % Normalize test set using training
stats
```

```
%% Step 4: Save splits and normalization parameters
save('Stratified_Splits_80_20.mat', 'X_train', 'Y_train', 'X_test',
'Y_test', 'mu', 'sigma');
```

Model Training and Testing (80/20 Split)

```
%% Step 1: Load stratified 80/20 data and normalization parameters
load('Stratified_Splits_80_20.mat', 'X_train', 'Y_train', 'X_test',
'Y_test', 'mu', 'sigma');

%% Step 2: Train linear regression model
model80 = fitlm(X_train, Y_train);

%% Step 3: Predict on test set
Y_pred = predict(model80, X_test);

%% Step 4: Save model and parameters
save('model_stratified_80.mat', 'model80', 'X_train', 'Y_train',
'X_test', 'Y_test', 'mu', 'sigma');
```

Generalization Test on New Pipes (Using 70/30 Model)

```
clear; clc;
%% Step 1: Load the saved 70/30 model and normalization parameters
load('model_70.mat', 'model70', 'mu', 'sigma');

%% Step 2: Load new test data
T_new = readtable('test_data.csv');
X_new = T_new{:, 1:15}; % New input features
Y_true = T_new{:, end}; % Actual outputs

%% Step 3: Normalize new data using training mean and std
X_new = (X_new - mu) ./ sigma;

%% Step 4: Predict using the loaded 70/30 model
Y_pred = predict(model70, X_new);
```

testing with new unseen pipes 80/20 modal:

```
clear; clc;
```

```
%% Step 1: Load the saved 80/20 model and normalization parameters
load('model_80.mat', 'model80', 'mu', 'sigma');

%% Step 2: Load new test data
T_new = readtable('test_data.csv');
X_new = T_new{:, 1:15}; % New input features
Y_true = T_new{:, end}; % Actual outputs

%% Step 3: Normalize new data using training mean and std
X_new = (X_new - mu) ./ sigma;

%% Step 4: Predict using the loaded 70/30 model
Y_pred = predict(model70, X_new);
```

Angular position 70/80 split:

```
%% Step 1: Load stratified 70/30 split with normalized data
load('Stratified_Splits_70_30.mat', 'X_train', 'Y_train', 'X_test',
'Y_test', 'mu', 'sigma');

%% Step 2: Load original data to replace labels with AngularPos
data = readmatrix('pipe_fault_signatures(22.csv');
AngularPos = data(:, end); % Assuming AngularPos is in the last column

% Reconstruct indices to match original split
n = size(AngularPos, 1);
trainIdx = false(n, 1);
testIdx = false(n, 1);
unique_labels = unique(AngularPos);
rng(42);

for i = 1:numel(unique_labels)
 label = unique_labels(i);
 indices = find(AngularPos == label);
 n_label = numel(indices);
 n_train = round(0.7 * n_label);
 shuffled = indices(randperm(n_label));
 trainIdx(shuffled(1:n_train)) = true;
 testIdx(shuffled(n_train+1:end)) = true;
end

Y_train = AngularPos(trainIdx);
Y_test = AngularPos(testIdx);
```

```
%% Step 3: Train model
model_angular = fitlm(X_train, Y_train);

%% Step 4: Predict
Y_train_pred = predict(model_angular, X_train);
Y_test_pred = predict(model_angular, X_test);

%% Step 5: Save the model
save('model_angular_70.mat', 'model_angular', 'X_train', 'Y_train',
'X_test', 'Y_test', 'mu', 'sigma');
```

References

1. Wong B, McCann JA (2021) Failure detection methods for pipeline networks: From acoustic sensing to cyber-physical systems†. Sensors 21(15):4959. https://doi.org/10.3390/S21154959
2. Tuninetti V, Huentemilla M, Gómez Á, Oñate A, Menacer B, Narayan S, Montalba C (2025) Evaluating pipeline inspection technologies for enhanced corrosion detection in mining water transport systems. Appl Sci (Switz) 15(3):1316. https://doi.org/10.3390/APP15031316
3. Korlapati NVS, Khan F, Noor Q, Mirza S, Vaddiraju S (2022) Review and analysis of pipeline leak detection methods. J Pipeline Sci Eng 2(4):100074. https://doi.org/10.1016/J.JPSE.2022.100074
4. Wu, H., Jiang, Z., Zhang, X., & Cheng, J. (2025). Research on a novel unsupervised-learning-based pipeline leak detection method based on temporal Kolmogorov–Arnold Network with autoencoder integration. Sensors, 25(2), 0384. https://doi.org/10.3390/s25020384
5. Patil S, Banerjee S, Tallur S (2024) Smart structural health monitoring (SHM) system for on-board localization of defects in pipes using torsional ultrasonic guided waves. Sci Rep 14(1): 24455. https://doi.org/10.1038/S41598-024-76236-W
6. Azanaw GM (2024) Integrating AI in structural health monitoring (SHM): a systematic review on ad-vances, challenges, and future directions. https://doi.org/10.2139/ssrn.5004977
7. Ravizza G, Ferrari R, Rizzi E, Dertimanis V, Chatzi EN (2019) Denoising corrupted structural vibration response: Critical comparison and assessment of related methods. COMPDYN Proceedings 3:3893–3904. https://doi.org/10.7712/120119.7194.19291
8. Malekloo A, Ozer E, AlHamaydeh M, Girolami M (2022) Machine learning and structural health monitoring overview with emerging technology and high-dimensional data source highlights. Struct Health Monit 21(4):1906–1955. https://doi.org/10.1177/14759217211036880
9. Mohamed A, Hamdi MS, Tahar S (2017) Using computational intelligence for the safety assessment of oil and gas pipelines: a survey. In: Studies in Big Data, vol 24. Springer Science and Business Media Deutschland GmbH, pp 189–207. https://doi.org/10.1007/978-3-319-53474-9_9
10. Pérez-Pérez E-J, González-Baldizón Y, Fragoso-Mandujano J-A, Guzmán-Rabasa J-A, Santos-Ruiz I (2024) Data-driven fault diagnosis in water pipelines based on neuro-Fuzzy Zonotopic Kalman filters. Math Comput Appl 30(1):2. https://doi.org/10.3390/mca30010002
11. Saidin SS, Jamadin A, Abdul Kudus S, Mohd Amin N, Anuar MA (2022) An overview: the application of vibration-based techniques in bridge structural health monitoring. Int J Concr Struct Mater 16(1):69. https://doi.org/10.1186/s40069-022-00557-1
12. Leite D, Andrade E, Rativa D, Maciel AMA (2025) Fault detection and diagnosis in industry 4.0: a review on challenges and opportunities. Sensors 25(1):60. https://doi.org/10.3390/s25010060

13. Singh P, Ahmad UF, Yadav S (2020) Structural health monitoring and damage detection through machine learning approaches. E3S Web of Conferences 220:01096. https://doi.org/10.1051/e3sconf/202022001096
14. El Dahr R, Lignos X, Papavieros S, Vayas I (2022) Design and validation of an accurate low-cost data acquisition system for structural health monitoring of a pedestrian bridge. J Civ Eng Constr 11(3):113–126. https://doi.org/10.32732/jcec.2022.11.3.113
15. Mustapha S, Lu Y, Ng CT, Malinowski P (2021) Sensor networks for structures health monitoring: placement, implementations, and challenges—a review. Vibration 4(3):551–585. https://doi.org/10.3390/vibration4030033
16. Jiang Y, Zhu L (2018) Modal analysis of liquid-filled pipeline under fluid-structure interaction by simulation and experiment methods. Vib Proce 21:42–47. https://doi.org/10.21595/vp.2018.20307
17. Lyu F, Zhou X, Ding Z, Qiao X, Song D (2024) Application research of ultrasonic-guided wave technology in pipeline corrosion defect detection: A review. Coatings 14(3):0358. https://doi.org/10.3390/coatings14030358
18. Han Y, Xia S, Wen Q, Zhang P, Lv F, Tang Z (2024) Nonuniform temperature compensation in ultrasonic guided wave pipeline health monitoring using local phase matching. Appl Sci (Switz) 14(22):10526. https://doi.org/10.3390/app142210526
19. Yu X, Fu Y, Li J, Mao J, Hoang T, Wang H (2024) Recent advances in wireless sensor networks for structural health monitoring of civil infrastructure. J Infrastruct Intell Resil 3(1):100066. https://doi.org/10.1016/j.iintel.2023.100066
20. Plevris V, Papazafeiropoulos G (2024) AI in structural health monitoring for infrastructure maintenance and safety. Infrastructures 9(12):0225. https://doi.org/10.3390/infrastructures9120225
21. Rabi RR, Vailati M, Monti G (2024) Effectiveness of vibration-based techniques for damage localization and lifetime prediction in structural health monitoring of bridges: a comprehensive review. Buildings 14(4):1183. https://doi.org/10.3390/buildings14041183
22. Kamiel BP, Rukmana I (2022) Vibration-based discriminant analysis for pipeline leaks detection. JMPM 6(2):8–15. https://doi.org/10.18196/JMPM.V6I2.16185
23. Chu T, Nguyen T, Yoo H, Wang J (2024) A review of vibration analysis and its applications. Heliyon 10(5). https://doi.org/10.1016/J.HELIYON.2024.E26282
24. Das S, Saha P, Patro SK (2016) Vibration-based damage detection techniques used for health monitoring of structures: a review. J Civ Struct Heal Monit 6(3):477–507. https://doi.org/10.1007/s13349-016-0168-5
25. Razi P, Taheri F (2014) A vibration-based strategy for health monitoring of offshore pipelines' girth-welds. Sensors (Switzerland) 14(9):17174–17191. https://doi.org/10.3390/S140917174
26. Golchinfar B, Ramezani MG, Donskoy D, Saboonchi H (2020) Vibro-Acoustic Modulation technique comparison with conventional nondestructive evaluation methods. www.mistrasgroup.com
27. Zajam S, Joshi T, Bhattacharya B (2019) Application of wavelet analysis and machine learning on vibration data from gas pipelines for structural health monitoring. Procedia Struct Integr 14:712–719. https://doi.org/10.1016/J.PROSTR.2019.05.089
28. Altabey WA, Noori M (2022) Artificial-intelligence-based methods for structural health monitoring. Appl Sci (Switz) 12(24):2726. https://doi.org/10.3390/app122412726
29. Pan H, Azimi M, Gui G, Yan F, Lin Z (2018) Vibration-based support vector machine for structural health monitoring. In: Lecture Notes in Civil Engineering, vol 5. Springer, pp 167–178. https://doi.org/10.1007/978-3-319-67443-8_14
30. Zhang T, Shi D, Wang Z, Zhang P, Wang S, Ding X (2022) Vibration-based structural damage detection via phase-based motion estimation using convolutional neural networks. Mech Syst Signal Process 178:109320. https://doi.org/10.1016/j.ymssp.2022.109320
31. Rajasekaran U, Kothandaraman M, Pua CH (2024) Water pipeline leak detection and localization with an integrated AI technique. IEEE Access 13:2736–2745. https://doi.org/10.1109/ACCESS.2024.3524669

32. Karabacak YE (2024) Condition monitoring of internal combustion engines with vibration signals and fault detection by using machine learning techniques. Int J Automot Eng Technol 13(4):191–200. https://doi.org/10.18245/ijaet.1251886
33. Samal PK, Sunil K, Jamadar IM, Srinidhi R (2024) AI-enhanced fault diagnosis in rolling element bearings: a comprehensive vibration analysis approach. FME Transactions 52(3): 450–460. https://doi.org/10.5937/fme2403450S
34. Barradas I, Garza LE, Morales-Menendez R, Vargas-Martínez A (2009) Leaks detection in a pipeline using artificial neural networks. https://doi.org/10.1007/978-3-642-10268-4_75
35. Al-Ammari WA, Sleiti AK (2024) Enhanced gas pipeline multiple leak detection using artificial neural networks in complex multiphase flow conditions. https://doi.org/10.46855/energy-proceedings-11220
36. Khan, A., Widdop, P. D., Day, A. J., Wood, A. S., Mounce, S. R., & Machell, J. (2008). Artificial neural network model for a low cost failure sensor: Performance assessment in pipeline distribution. International Journal of Mathematical, Computational, Physical, Electrical and Computer Engineering, 2(9), 690–696.
37. Deng F, Tao X, Wei P, Wei S (2023) A robust deep learning-based damage identification approach for SHM considering missing data. Appl Sci 13(9):5421
38. Bagri I, Tahiry K, Hraiba A, Touil A, Mousrij A (2024) Vibration signal analysis for intelligent rotating machinery diagnosis and prognosis: a comprehensive systematic literature review. Vibration 7(4):1013–1062. https://doi.org/10.3390/VIBRATION7040054
39. Zuo J, Deng X, Sutton MA (2005) Advances in tetrahedral mesh generation for modelling of three-dimensional regions with complex, curvilinear crack shapes. Int J Numer Methods Eng 63(3):256–275
40. Lengvarský P, Hagara M, Schrötter M, Bocko J (2014) Modal analysis of axially symmetric structure. Am J Mech Eng 2(7):265–269. http://pubs.sciepub.com/ajme/2/7/19
41. Craig RR, Kurdila AJ (2006) Fundamentals of structural dynamics, 2nd edn. Wiley, Hoboken, NJ
42. ANSYS, Inc (2021) ANSYS mechanical APDL theory reference manual. Release 2021 R2. ANSYS, Inc., Canonsburg, PA
43. Bai Z, Demmel J, Dongarra J, Ruhe A, van der Vorst H (2000) Templates for the solution of algebraic eigenvalue problems: a practical guide. SIAM, Philadelphia, PA
44. Mohammadzaheri M, Amouzadeh A, Doustmohammadi M, Emadi M, Jamshidi E, Ghodsi M, Soltani P (2021) Fuzzy analysis of resonance frequencies for structural inspection of an engine cylinder block. Fuzzy Inf Eng 13(3):266–276. https://doi.org/10.1080/16168658.2021.1908819
45. LeCun Y, Bottou L, Orr GB, Müller KR (2012) Efficient backprop. In: Montavon G, Orr GB, Müller KR (eds) Neural networks: Tricks of the trade. Springer, pp 9–48. https://doi.org/10.1007/978-3-642-35289-8_3
46. Hornik K, Stinchcombe M, White H (1989) Multilayer feedforward networks are universal approximators. Neural Netw 2(5):359–366. https://doi.org/10.1016/0893-6080(89)90020-8
47. Kessai I, Benammar S, Doghmane MZ, Tee KF (2022) Estimation of circular arc crack depths and locations in rotary drilling pipes subjected to free vibrations. Vibration 5(1):11. https://doi.org/10.3390/vibration5010011
48. Saha P, Giurgiutiu V (2023) A neural network approach to estimate the frequency of a cantilever beam with random multiple damage. Sensors 23(19):7867. https://doi.org/10.3390/s23197867
49. Pettersson K, Karzhou A, Pettersson I (2020) A feedforward neural network for modelling of average pressure frequency response. arXiv preprint arXiv:2012.02276. https://arxiv.org/abs/2012.02276
50. Kourehli SS (2024) Structural damage detection under short time load using cascade-forward network. J Rehabil Civ Eng 12(3):32–42. https://civiljournal.semnan.ac.ir/article_8403.html
51. Mohammadzaheri H, Zhou J, Al-Jumaily A (2011) Artificial neural networks for modeling and control of a liquid level system: A comparative study. J Intell Robot Syst 64(3–4):463–476. https://doi.org/10.1007/s10846-011-9564-7

52. Hinton GE, Osindero S, Teh YW (2006) A fast learning algorithm for deep belief nets. Neural Comput 18(7):1527–1554. https://doi.org/10.1162/neco.2006.18.7.1527
53. Stathakis D (2009) How many hidden layers and nodes? Int J Remote Sens 30(8):2133–2147
54. Xie, Y. G., Gao, C. G., Wang, P. Z., Zhou, L., Zhang, C. J., & Qu, X. Q. (2023). Research on vibration fatigue damage locations of offshore oil and gas pipelines based on the GA-improved BP neural network. Shock Vib, 2023, 2530651. https://doi.org/10.1155/2023/2530651
55. Hagan MT, Menhaj MB (1994) Training feedforward networks with the Marquardt algorithm. IEEE Trans Neural Netw 5(6):989–993
56. Demuth H, Beale M (1999) Neural network design. PWS Publishing
57. Ripley BD (2008) Pattern recognition and neural networks. Cambridge University Press
58. Reitermanová Z (2010) Data splitting. In: WDS'10 proceedings of contributed papers, Part I. MATFYZPRESS, pp 31–36
59. Yan R, Wang S (2022) Linear regression models. In: Applications of machine learning and data analytics models in maritime transportation. IET, pp 51–62. https://doi.org/10.1049/PBTR038E_CH5
60. Chopra, A. K. (2017). Dynamics of Structures: Theory and Applications to Earthquake Engineering (5th ed.). Hoboken, NJ: Pearson Education. ISBN: 978-0134555126
61. Laabid, Z., Moumen, A., Mansouri, K., & Siadat, A. (2023). Numerical study of the speed's response of the various intelligent models using the tansig, logsig and purelin activation functions in different layers of artificial neural network. IAES International Journal of Artificial Intelligence (IJ-AI), 12(1), 155–161. https://doi.org/10.11591/ijai.v12.i1.pp155-161
62. Sáez, J. A., & Romero-Béjar, J. L. (2022). On the suitability of bagging-based ensembles with borderline label noise. Mathematics, 10(11), 1892. https://doi.org/10.3390/math10111892
63. Hastie, T., Tibshirani, R., & Friedman, J. (2001). The Elements of Statistical Learning: Data Mining, Inference, and Prediction. New York: Springer. ISBN: 978-0387952840

MIX
Papier aus verantwortungsvollen Quellen
Paper from responsible sources
FSC® C105338

If you have any concerns about our products,
you can contact us on
ProductSafety@springernature.com

In case Publisher is established outside the EU,
the EU authorized representative is:
Springer Nature Customer Service Center GmbH
Europaplatz 3, 69115 Heidelberg, Germany

Printed by Libri Plureos GmbH
in Hamburg, Germany